T0255071

Lecture Notes in Computer Science 13730

More information about this series at https://link.springer.com/bookseries/558

Bo Hu · Yunni Xia · Yiwen Zhang ·
Liang-Jie Zhang (Eds.)

Big Data – BigData 2022

11th International Conference
Held as Part of the Services Conference Federation, SCF 2022
Honolulu, HI, USA, December 10–14, 2022
Proceedings

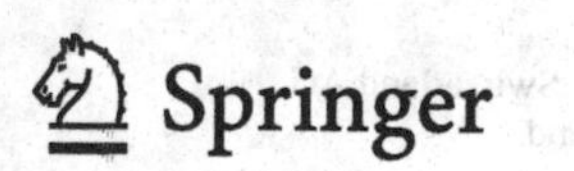

Editors
Bo Hu
Shenzhen Yihuo Technology Co., Ltd.
Shenzhen, China

Yunni Xia
Chongqing University
Chongqing, China

Yiwen Zhang
Anhui University
Hefei, China

Liang-Jie Zhang
Kingdee International Software
Group Co., Ltd.
Shenzhen, China

ISSN 0302-9743 ISSN 1611-3349 (electronic)
Lecture Notes in Computer Science
ISBN 978-3-031-23500-9 ISBN 978-3-031-23501-6 (eBook)
https://doi.org/10.1007/978-3-031-23501-6

This Springer imprint is published by the registered company Springer Nature Switzerland AG
The registered company address is: Gewerbestrasse 11, 6330 Cham, Switzerland

Preface

The 2022 International Congress on Big Data (BigData 2022) provided an international forum to formally explore business insights of all kinds of value-added "services". Big Data is a key enabler of exploring business insights and the economics of services.

BigData 2022 was one of the events of the Services Conference Federation event (SCF 2022), which had the following 10 collocated service-oriented sister conferences: the International Conference on Web Services (ICWS 2022), the International Conference on Cloud Computing (CLOUD 2022), the International Conference on Services Computing (SCC 2022), the International Conference on Big Data (BigData 2022), the International Conference on AI & Mobile Services (AIMS 2022), the International Conference on Metaverse (METAVERSE 2022), the International Conference on Internet of Things (ICIOT 2022), the International Conference on Cognitive Computing (ICCC 2022), the International Conference on Edge Computing (EDGE 2022), and the International Conference on Blockchain (ICBC 2022).

This volume presents the papers accepted at BigData 2022. Its major topics included Big Data Architecture, Big Data Modeling, Big Data as a Service, Big Data for Vertical Industries (Government, Healthcare, etc.), Big Data Analytics, Big Data Toolkits, Big Data Open Platforms, Economic Analysis, Big Data for Enterprise Transformation, Big Data in Business Performance Management, Big Data for Business Model Innovations and Analytics, Big Data in Enterprise Management Models and Practices, Big Data in Government Management Models and Practices, and Big Data in Smart Planet Solutions.

We received 16 submissions and accepted 9 papers. Each was reviewed and selected by at least three independent members of the Program Committee. We are pleased to thank the authors whose submissions and participation made this conference possible. We also want to express our thanks to the Organizing Committee and Program Committee members for their dedication in helping to organize the conference and review the submissions.

December 2022

Bo Hu
Yunni Xia
Yiwen Zhang
Liang-Jie Zhang

Preface

The 2022 International Conference on Big Data (BigData 2022) [illegible] international forum [illegible] services. Big Data is a [illegible]

BigData 2022 [illegible] 2022 [illegible] the International Conference on [illegible] Services [illegible] 2022, the International Conference on Cloud Computing (CLOUD 2022), the International Conference on [illegible] (BigData [illegible]), the International Conference on [illegible] (ICWS 2022), the International Conference on [illegible] (SERVICES 2022), the International Conference [illegible] 2022, the International Conference on Cognitive Computing (ICCC 2022), the International Conference on [illegible] 2022, and the International Conference on Blockchain (ICBC 2022).

This volume presents the papers [illegible] BigData 2022 [illegible] Big Data [illegible] Big Data [illegible] Data Platform [illegible] Enterprise Transformation [illegible] Business [illegible] Management, [illegible] and [illegible] Big Data [illegible] Smart Planet [illegible]

[illegible] papers [illegible] by at least [illegible] We are pleased to thank the [illegible] We would also like to express our thanks [illegible] Program Committee members for their [illegible] in helping to [illegible] submissions.

December 2022

Organization

Services Conference Federation (SCF 2022)

General Chairs

Ali Arsanjani	Google, USA
Wu Chou	Essenlix, USA

Coordinating Program Chair

Liang-Jie Zhang	Kingdee International Software Group, China

CFO and International Affairs Chair

Min Luo	Georgia Tech, USA

Operation Committee

Jing Zeng	China Gridcom, China
Yishuang Ning	Tsinghua University, China
Sheng He	Tsinghua University, China

Steering Committee

Calton Pu	Georgia Tech, USA
Liang-Jie Zhang	Kingdee International Software Group, China

Bigdata 2022

General Chair

James Wang	DAMA, China

Program Chairs

Bo Hu	Shenzhen Yihuo Technology Co., Ltd., China
Yunni Xia	Chongqing University, China
Yiwen Zhang	Anhui University, China

Program Committee

Guobing Zou	Shanghai University, China
Ying Chen	Beijing Information Science and Technology University, China
Zhihui Lv	Fudan University, China
Yutao Ma	Wuhan University, China
Wuhui Chen	Sun Yat-Sen University, China
Kai Peng	Huaqiao University, China
Hui Dou	Anhui University, China
Peter Baumann	Jacobs University, Germany
Shreyansh Bhatt	Amazon, USA
Marios Dikaiakos	University of Cyprus, Cyprus
Jinzhu Gao	University of the Pacific, USA
Anastasios Gounaris	Aristotle University of Thessaloniki, Greece
Verena Kantere	University of Ottawa, Canada
Sam Supakkul	NCR Corporation, USA
Nan Wang	Heilongjiang University, China
Wenbo Wang	GoDaddy, USA
Daqing Yun	Harrisburg University, USA

Services Society

The Services Society (S2) is a non-profit professional organization that was created to promote worldwide research and technical collaboration in services innovations among academia and industrial professionals. Its members are volunteers from industry and academia with common interests. S2 is registered in the USA as a "501(c) organization", which means that it is an American tax-exempt nonprofit organization. S2 collaborates with other professional organizations to sponsor or co-sponsor conferences and to promote an effective services curriculum in colleges and universities. S2 initiates and promotes a "Services University" program worldwide to bridge the gap between industrial needs and university instruction.

The Services Society has formed Special Interest Groups (SIGs) to support technology- and domain-specific professional activities:

- Special Interest Group on Web Services (SIG-WS)
- Special Interest Group on Services Computing (SIG-SC)
- Special Interest Group on Services Industry (SIG-SI)
- Special Interest Group on Big Data (SIG-BD)
- Special Interest Group on Cloud Computing (SIG-CLOUD)
- Special Interest Group on Artificial Intelligence (SIG-AI)
- Special Interest Group on Edge Computing (SIG-EC)
- Special Interest Group on Cognitive Computing (SIG-CC)
- Special Interest Group on Blockchain (SIG-BC)
- Special Interest Group on Internet of Things (SIG-IOT)
- Special Interest Group on Metaverse (SIG-Metaverse)

Services Conference Federation (SCF)

As the founding member of SCF, the first International Conference on Web Services (ICWS) was held in June 2003 in Las Vegas, USA. The First International Conference on Web Services - Europe 2003 (ICWS-Europe'03) was held in Germany in October 2003. ICWS-Europe'03 was an extended event of the 2003 International Conference on Web Services (ICWS 2003) in Europe. In 2004 ICWS-Europe changed to the European Conference on Web Services (ECOWS), which was held in Erfurt, Germany.

SCF 2019 was held successfully during June 25–30, 2019 in San Diego, USA. Affected by COVID-19, SCF 2020 was held online successfully during September 18–20, 2020, and SCF 2021 was held virtually during December 10–14, 2021.

Celebrating its 20-year birthday, the 2022 Services Conference Federation (SCF 2022, www.icws.org) was a hybrid conference with a physical onsite in Honolulu, Hawaii, USA, satellite sessions in Shenzhen, Guangdong, China, and also online sessions for those who could not attend onsite. All virtual conference presentations were given via prerecorded videos in December 10–14, 2022 through the BigMarker Video Broadcasting Platform: https://www.bigmarker.com/series/services-conference-federati/series_summit.

Just like SCF 2022, SCF 2023 will most likely be a hybrid conference with physical onsite and virtual sessions online, it will be held in September 2023.

To present a new format and to improve the impact of the conference, we are also planning an Automatic Webinar which will be presented by experts in various fields. All the invited talks will be given via prerecorded videos and will be broadcast in a live-like format recursively by two session channels during the conference period. Each invited talk will be converted into an on-demand webinar right after the conference.

In the past 19 years, the ICWS community has expanded from Web engineering innovations to scientific research for the whole services industry. Service delivery platforms have been expanded to mobile platforms, the Internet of Things, cloud computing, and edge computing. The services ecosystem has been enabled gradually, with value added and intelligence embedded through enabling technologies such as Big Data, artificial intelligence, and cognitive computing. In the coming years, all transactions involving multiple parties will be transformed to blockchain.

Based on technology trends and best practices in the field, the Services Conference Federation (SCF) will continue to serve as a forum for all services-related conferences. SCF 2022 defined the future of the new ABCDE (AI, Blockchain, Cloud, Big Data & IOT). We are very proud to announce that SCF 2023's 10 colocated theme topic conferences will all center around "services", while each will focus on exploring different themes (Web-based services, cloud-based services, Big Data-based services, services innovation lifecycles, AI-driven ubiquitous services, blockchain-driven trust service ecosystems, Metaverse services and applications, and emerging service-oriented technologies).

The 10 colocated SCF 2023 conferences will be sponsored by the Services Society, the world-leading not-for-profit organization dedicated to serving more than 30,000 services computing researchers and practitioners worldwide. A bigger platform means bigger opportunities for all volunteers, authors, and participants. Meanwhile, Springer will provide sponsorship for Best Paper Awards. All 10 conference proceedings of SCF 2023 will be published by Springer, and to date the SCF proceedings have been indexed in the ISI Conference Proceedings Citation Index (included in the Web of Science), the Engineering Index EI (Compendex and Inspec databases), DBLP, Google Scholar, IO-Port, MathSciNet, Scopus, and ZbMath.

SCF 2023 will continue to leverage the invented Conference Blockchain Model (CBM) to innovate the organizing practices for all 10 conferences. Senior researchers in the field are welcome to submit proposals to serve as CBM ambassadors for individual conferences.

SCF 2023 Events

The 2023 edition of the Services Conference Federation (SCF) will include 10 service-oriented conferences: ICWS, CLOUD, SCC, BigData, AIMS, METAVERSE, ICIOT, EDGE, ICCC and ICBC.

The 2023 International Conference on Web Services (ICWS 2023, http://icws.org/2023) will be the flagship theme-topic conference for Web-centric services, enabling technologies and applications.

The 2023 International Conference on Cloud Computing (CLOUD 2023, http://thecloudcomputing.org/2023) will be the flagship theme-topic conference for resource sharing, utility-like usage models, IaaS, PaaS, and SaaS.

The 2023 International Conference on Big Data (BigData 2023, http://bigdatacongress.org/2023) will be the theme-topic conference for data sourcing, data processing, data analysis, data-driven decision-making, and data-centric applications.

The 2023 International Conference on Services Computing (SCC 2023, http://thescc.org/2023) will be the flagship theme-topic conference for leveraging the latest computing technologies to design, develop, deploy, operate, manage, modernize, and redesign business services.

The 2023 International Conference on AI & Mobile Services (AIMS 2023, http://ai1000.org/2023) will be a theme-topic conference for artificial intelligence, neural networks, machine learning, training data sets, AI scenarios, AI delivery channels, and AI supporting infrastructures, as well as mobile Internet services. AIMS will bring AI to mobile devices and other channels.

The 2023 International Conference on Metaverse (Metaverse 2023, http://Metaverse1000.org) will focus on innovations of the services industry, including financial services, education services, transportation services, energy services, government services, manufacturing services, consulting services, and other industry services.

The 2023 International Conference on Cognitive Computing (ICCC 2023, http://thecognitivecomputing.org/2023) will focus on leveraging the latest computing technologies to simulate, model, implement, and realize cognitive sensing and brain operating systems.

The 2023 International Conference on Internet of Things (ICIOT 2023, http://iciot.org/2023) will focus on the science, technology, and applications of IOT device innovations as well as IOT services in various solution scenarios.

The 2023 International Conference on Edge Computing (EDGE 2023, http://theedgecomputing.org/2023) will be a theme-topic conference for leveraging the latest computing technologies to enable localized device connections, edge gateways, edge applications, edge-cloud interactions, edge-user experiences, and edge business models.

The 2023 International Conference on Blockchain (ICBC 2023, http://blockchain1000.org/2023) will concentrate on all aspects of blockchain, including digital currencies, distributed application development, industry-specific blockchains, public blockchains, community blockchains, private blockchains, blockchain-based services, and enabling technologies.

Contents

Contents

A Massive Data Retrieval Method for Power Information Collection Systems

Xiang Wang[1], Haimin Hong[1,3], Zhanxia Wu[1,2,3], Jing Zeng[1](✉), and Guochuan Liu[1]

[1] Department of Research Center, China Gridcom Co., Ltd., Shenzhen, China
jerryzengjing@163.com

[2] Department of Research Center, Beijing Smart-Chip Microelectronics Technology Co., Ltd., Beijing, China

[3] Department of Research Center, Shenzhen Smart-Chip Microelectronics Technology Co., Ltd., Shenzhen, China

Abstract. With the development of smart grid and big data, massive data retrieval has received intensive attentions by practitioners and developers, especially for its using in power information collection applications. It is a tricky task for improving the retrieval efficiency of massive data. The main big data solutions are likely to store the data in Hbase, which is difficult to improve the performance when querying small data from a large volume of data. Existing methods for this problem are inclined to use phoenix to query Hbase, but they exhibit inefficient under multiple table joining. In this paper, to sovle the quick retrieval for massive data in power information collection systems, we present a method via using Elasticsearch and Hbase to augment the performance of massive data retrieval. Specifically, we store the Rowkey and index information in elasticsearch to help accelerate the querying of Hbase. Then an experiment in real use case is used to verify the effectiveness of presented method.

Keywords: Bigdata · Massive data retrieval · Power information collection

1 Introduction

With the development of big data and digital transformation of power grid [1], massive data retrieval is becoming an essential part in power information collection systems [7]. Currently, massive data storage is mainly implemented based on NoSQL storage components and HDFS. As a full data storage medium, HDFS can only be used for data archiving and cannot support fast data retrieval due to its data storage characteristics. In addition to provide massive data storage, NoSQL storage to a certain extent can also support random retrieval of massive data. However, compared with traditional relational storage, NoSQL storage still cannot meet the requirements of data retrieval efficiency. Through analysis, NoSQL storage database has the following advantages: 1) It has a very high load

B. Hu et al. (Eds.): BigData 2022, LNCS 13730, pp. 1–8, 2022.
https://doi.org/10.1007/978-3-031-23501-6_1

speed, up to the sum of disk I/O in the cluster; 2) It is capable of storing massive data. Horizontal expansion of the storage capability is simple; 3) It supports real-time data loading in massive data scenarios; 4) It supports efficient compression, for saving disk storage, memory and CPU resource. However, it also has mutiple disadvantages. The efficiency of retrieving a small amount of data is quite low. Random data retrieval, update and other scenarios are inefficient. Efficiency in batch update scenarios heavily depends on the design of storage architecture. It is not suitable for real-time deletion and update scenarios.

Some projects in power information collection systems use Kudu and HBase [11] components in real application environment, the two types of storage can only support the batch computing scenario of mass data, but are unable to meet the needs of the main applications of the resulting data query. The main reason is that the column type several positions cannot be as flexible as a traditional RDBMS to indexing. If the query fails to match the Rowkey or Kudu primary key of HBase, the query is changed to full table scan, resulting in slow query and waste of cluster computing resources. For example, in HBase, the Rowkey is the electricity meter ID. We need to find all electricity meters in the same unit based on the power supply unit. If the Rowkey does not use the power supply unit as the start code of the Rowkey, it will scan all rowkeys using either the RowFilter or the ColumnFilter. It is slow and causes I/O overload in the Hbase cluster. If the power supply unit is set to the Rowkey start code, only the power supply unit can be used to search for the power supply unit. In addition, an excessively long Rowkey may adversely affect HBase storage and efficiency. Existing solutions used storage components that support fast data retrieval to create secondary indexes for HBase and Kudu, they query the index database to quickly find the Rowkey or primary key, and use them to get data and bypass the scan process.

Most of current solutions for power information collection systems [8] have used Oracle for Rowkey query splicing. But these solutions will bring extra burden to Oracle storage and are not suitable for a large number of secondary index business scenarios. They are only suitable for a small amount of data scenarios. In this paper, we present an Elasticsearch (ES) with HBase solution for massive data rapid retrieval technology, to solve that the current column data warehouse is not suitable for scanning a small amount of data, resulting in data sharing difficulties with external systems. The key idea is to use ES to save the index including Rowkey, metering information, measuring point, etc. When we proceed the retrieval for massive data in Hbase, a quick querying for index is firstly conducted by using Elasticsearch to help improve the querying efficiency of Hbase.

The reminder of this paper is organized as follows: Sect. 2 gives the related works about massive data retrieval. In Sect. 3, we present a method for massive data retrieval in detail. Then, an experiment is carried out and demonstrated in Sect. 4. Finally, the conclusions of the paper are summarized in Sect. 5.

2 Related Works

Data in power information collection systems [5] is classified into structured data and unstructured data:

1) Structured data: it is also known as row data, which is logically expressed and realized by two-dimensional table structure, strictly follows data format and length specifications, and is mainly stored and managed by relational database. It refers to data of fixed format or limited length, such as database, metadata, etc.

2) Unstructured data: it is also known as full-text data, variable length or no fixed format, not suitable for two-dimensional table performance by the database, including all formats of office documents, XML, HTML, Word documents, emails, all kinds of reports, pictures and AUDIos, video information, etc. Note: XML and HTML can be classified as semi-structured data if we want to be more specific. Because they also have their own specific tag formats, they can be processed as structured data or extract plain text as unstructured data as needed.

According to two kinds of data classification, search is also divided into two kinds: structured data search and unstructured data search. For structured data, data has a specific structure and is generally stored and searched in the form of two-dimensional tables in relational databases (Mysql,Oracle,etc.). For unstructured data, there are two main methods [4] to search full-text data: sequential scanning and full-text retrieval.

1) Sequential scan: Search specific keywords in sequential scan mode. For example, search a newspaper and find out where the words containing peace appeared in the newspaper. In this case, it is necessary to scan the contents of the newspaper from beginning to end and then mark all the keywords. This scanning method is inefficient, and the scanning time is proportional to the amount of data.

2) Full-text retrieval: Part of the information in unstructured data is extracted and reorganized to make it have a certain structure. On this basis, the data is searched. The information extracted from unstructured data and then reorganized is called index. In this way, the main work is to create the index in the early stage. But after the index is completed, the data retrieval efficiency can be effectively improved.

Existing massive data retrieval for smart grid divided into two types: Phoenix combined with Hbase and cloud vendors based. For Phoenix plus Hbase [2, 6,9], it uses the Phoenix query engine to operate Hbase data. The solution has the following advantages: (1) Phoenix supports structured query Language (SQL), which simplifies the difficulty of accessing Hbase data. Its disadvantages are summarized as follow: (1) Phoenix components are heavy, relatively high difficulty in operation and maintenance; (2) Phoenix still has some problems in terms of performance, and does not realize perfect index push-down. In many cases, the index query needs to initiate two interactions between the client and the server. The first time to obtain the matched index information, and the second time is the matched data. (3) The scheme has been applied in the project.

The actual application shows that the efficiency of the scheme is good for Hbase single table data query, but low for complex multi-table associated query.

HBase is a standard configuration of cloud services. Some public clouds also provide the secondary index function [3,10,12]. The advantages of this scheme are the one-stop service of the manufacturer, and the operation and maintenance cost is very low. The shortcomings are that mainstream cloud vendors encapsulate the combination of ES and HBase at the bottom of their technical solutions. Although interface encapsulation reduces the difficulty for development, performance loss caused by the architecture still exists.

Existing methods can not meet the general demands and implementation for power information collection systems. To achieve the goal for quickly retrieving massive power information, we use Elasticsearch for Hbase, which sharply reduce the access latency of Hbase.

3 The Proposed Method for Massive Data Retrieval

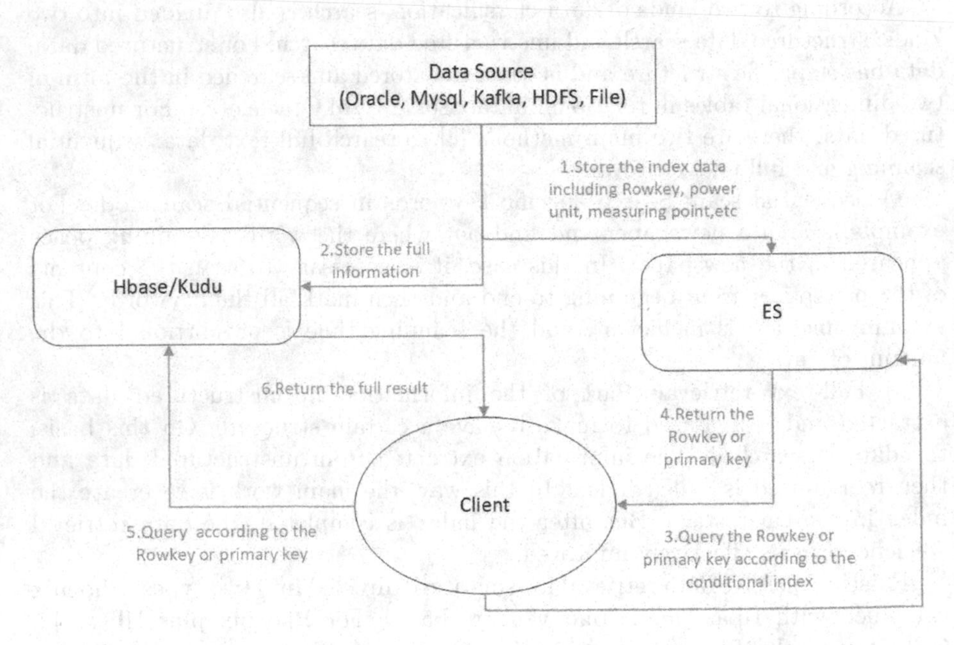

Fig. 1. The method of quick retrival for massive data in power information system

3.1 The Proposed Method

There are two types of technical combination for ES with Hbase. We need to choose the method based on actual service scenarios. For the first one, in service scenarios, it requires high data writing performance, data must be written to Hbase firstly and then to ES. The two written processes are independent of each other. This method maximizes the writing performance. The disadvantage is that data inconsistency may exist. For the second method, it uses the collaboration of Hbase components to process changes in listening data within Hbase. When data changes, indexes in ES are updated in real time. This method is a popular solution, but the co-processor affects the overall Hbase performance.

To make comparison more reliable, Spark is used for querying with the same entry and querying method. Based on Hadoop enabled big data platform, this study aims to solve practical problems within power data retrieval, and to carry out feasibility and effectiveness research of secondary index mode.

As shown in the Fig. 1, when data sources(Oracle,Mysql,Kafka, HDFS,File) are connected to the big data platform, HBase and ES are written at the same time. The full data is stored in HBase/Kudu for data analysis or sharing, and possible conditions for data retrieval are stored in ES. The Rowkey of power data usually is power unit, measuring point, etc., which is used to index the power data. The efficiency of the search engine (ES) is utilized to quickly retrieve HBase rowkeys. When an external system client accesses data, it firstly accesses the ES component, quickly retrieves the target Rowkey, and then accurately obtains the target data from HBase based on the Rowkey for display or analysis.

3.2 Pseudocode

As shown in Fig. 2, the process for massive data retrieval via ES with Hbase is demonstrated with pesudocode. Line 1–6 represents the initialized configuration for Hbase. In Line 7–13, it denotes the Rowkey is queryed from ES via Meter ID of power data table, the Rowkey can be used to accelerate the querying performance of Hbase. For Line 14–26, the selected data are precisely chosen from the Hbase for use.

4 The Experiment

In this technical verification scenario, we use power information data as a use case,51 pieces of data of a specific power supply unit are retrieved from 4 million pieces of power data. The configuration of Hadoop cluster is shown in Table 1.

For the big data computing engine, we choose Spark to conduct the task with same configuration. The running results are shown in Figs. 3 and 4. The traditional HBase searching time is 96 s, and the secondary index search method based on the ES component is 3.2 s, which is 30 times faster than the traditional method. From the experiment comparison, we found that the proposed method significantly outperformed traditional retrieval method for Hbase. The results also prove that it is quite feasible and efficient in the context of a large batch of data querying for Hbase.

```
1 val myConf = HBaseUtil.initHbase()
2 val hconn = ConnectionFactory.createConnection(myConf)
3 val myTable = hconn.getTable(TableName.valueOf("cj:es"))
4 val col = "MPED_ID, CP_NO, MPED_INDEX, MPED_NAME, METER_ID, CONS_ID, COLL_OBJ_ID, MP_ID,
5BELONG_FLAG, FMR_NO, ORG_NO".split(", ")
6val itr: Iterator[String] = p.map({ r=>{
7  //是否存在 rowkey 查询结果
8  if(!r.isNullAt(r.fieldIndex("METER_ID"))){
9    val rk = r.getAs[String]("METER_ID")
10     val get = new Get(rk.getBytes())
11    val result = myTable.get(get)
12     val json = new JSONObject()
13     //判断 hbase 中是否有数据如果有则装入
14      col.foreach(c =>{
15        if(!result.isEmpty &&result.containsColumn("f1".getBytes(), c.getBytes)){
16          json.put(c, Bytes.toString(result.getValue("f1".getBytes(), c.getBytes())))
17        }else{
18          json.put(c, null)
19        }
20      })
21     json.toJSONString
22  }else{
23     null
24  }
25 }
26 })
```

Fig. 2. The pseudocode of fast retrieval based on ES for Hbase

Table 1. The configuration of the Hadoop cluster for experiment.

Configuration	Description
Cluster Node Number	8
Disk	>=1T
CPU core	32 Cores, 64 Threads
Memory	64 G
OS	Centos 7.6
Bandwidth	1000 M

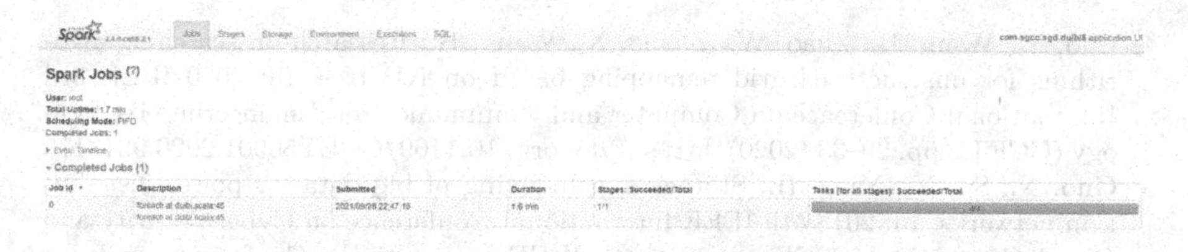

Fig. 3. The traditional querying and retrieval for Hbase

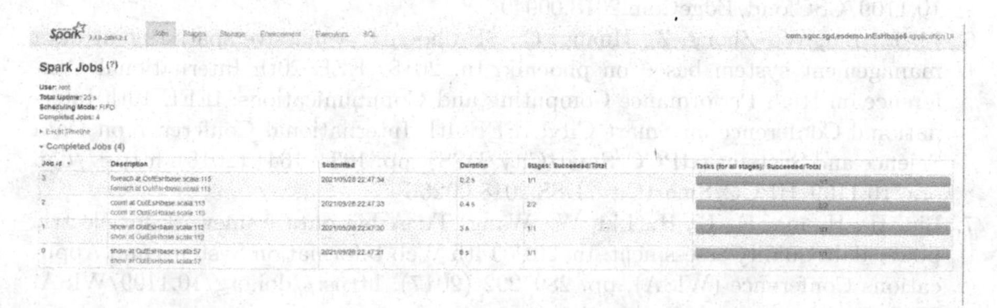

Fig. 4. The querying and retrieval based ES secondary index for HBase

5 Conclusions

In this paper, we present a quick retrieval method for massive data in power information collection system. Specifically, we leverage ES to proceed the optimization for Hbase querying. The Rowkey or index can be quick query via ES retrieval ability. The performance of the proposed method is verified through a case in real application scenario. The result has shown that the proposed method can significantly improves the accessing performance for Hbase under the massive data context. It can be promoted and widely applied in power data querying scenarios in future.

References

1. Abe, R., Taoka, H., McQuilkin, D.: Digital grid: communicative electrical grids of the future. IEEE Trans. Smart Grid **2**(2), 399–410 (2011). https://doi.org/10.1109/TSG.2011.2132744
2. Al-Badarneh, A., Najadat, H., Al-Soud, M., Mosaid, R.: Phoenix: a mapReduce implementation with new enhancements. In: 2016 7th International Conference on Computer Science and Information Technology (CSIT), pp. 1–5 (2016). https://doi.org/10.1109/CSIT.2016.7549451
3. Cao, C., Wang, W., Zhang, Y., Lu, J.: Embedding index maintenance in store routines to accelerate secondary index building in HBase. In: 2018 IEEE 11th International Conference on Cloud Computing (CLOUD), pp. 500–507 (2018). https://doi.org/10.1109/CLOUD.2018.00070

4. Cao, Y., Wang, B., Zhao, W., Zhang, X., Wang, H.: Research on searching algorithms for unstructured grid remapping based on KD tree. In: 2020 IEEE 3rd International Conference on Computer and Communication Engineering Technology (CCET), pp. 29–33 (2020). https://doi.org/10.1109/CCET50901.2020.9213175
5. Guo, N., Su, Y., Yang, H.: Storage and indexing of big data for power distribution networks. In: 2018 5th IEEE International Conference on Cyber Security and Cloud Computing (CSCloud)/2018 4th IEEE International Conference on Edge Computing and Scalable Cloud (EdgeCom), pp. 224–2243 (2018). https://doi.org/10.1109/CSCloud/EdgeCom.2018.00049
6. Li, L., Liu, W., Zhong, Z., Huang, C.: SP-phoenix: a massive spatial point data management system based on phoenix. In: 2018 IEEE 20th International Conference on High Performance Computing and Communications; IEEE 16th International Conference on Smart City; IEEE 4th International Conference on Data Science and Systems (HPCC/SmartCity/DSS), pp. 1634–1641 (2018). https://doi.org/10.1109/HPCC/SmartCity/DSS.2018.00266
7. Liu, H., Huang, F., Li, H., Liu, W., Wang, T.: A big data framework for electric power data quality assessment. In: 2017 14th Web Information Systems and Applications Conference (WISA), pp. 289–292 (2017). https://doi.org/10.1109/WISA.2017.29
8. Rongrong, S., Qing, L., Xin, S., Baifeng, N., Qiang, W.: Application of big data in power system reform. In: 2021 IEEE Asia-Pacific Conference on Image Processing, Electronics and Computers (IPEC), pp. 1340–1342 (2021). https://doi.org/10.1109/IPEC51340.2021.9421337
9. Touloupas, G., Konstantinou, I., Koziris, N.: Rasp: real-time network analytics with distributed nosql stream processing. In: 2017 IEEE International Conference on Big Data (Big Data), pp. 2414–2419 (2017). https://doi.org/10.1109/BigData.2017.8258198
10. Wu, H., et al.: A performance-improved and storage-efficient secondary index for big data processing. In: 2017 IEEE International Conference on Smart Cloud (SmartCloud), pp. 161–167 (2017). https://doi.org/10.1109/SmartCloud.2017.32
11. Zhao, F., Wang, G., Deng, C., Zhao, Y.: A real-time intelligent abnormity diagnosis platform in electric power system. In: 16th International Conference on Advanced Communication Technology, pp. 83–87 (2014). https://doi.org/10.1109/ICACT.2014.6778926
12. Zhu, Y., Xu, Q., Shi, H., Samsudin, J.: DS-index: a distributed search solution for federated cloud. In: 2016 IEEE International Conference on Networking, Architecture and Storage (NAS), pp. 1–4 (2016). https://doi.org/10.1109/NAS.2016.7549397

Research and Application of the Data Resource Directory System of the Aerospace Enterprise

Yuhan Ma(✉), Cheng Ji, Tingwei Fei, Zhengxuan Duan, and Jianchao Luo

Beijing Jinghang Research Institute of Computing and Communication, Beijing 100074, People's Republic of China
airma2020@163.com

Abstract. The data resource of the aerospace enterprise have the problems of scattered data, multiple sources, inconsistent data, numerous interfaces, and inconsistent standard rules. It is necessary to carry out data governance work systematically. In order to improve the standardization of data governance work such as data classification, data sharing and data accountability of the aerospace enterprise, we research on the data resource directory system of the aerospace enterprise. We focus on the research of data resource directory standard, data resource inventory, and data resource directory framework. We introduce the application practice of the data resource directory system of the aerospace enterprise, which verifies the effectiveness and practicability of this system.

Keywords: Data governance · Data resource directory · Data resource inventory · Aerospace enterprise

1 Introduction

With the proposal of the State Council of China in 2020 to build a more complete market based allocation system for factors, data is seem as "the five elements" with land, labor, capital and technology. The data factor market [1] has received more and more attention. In order to promote the development of the data element market, the State Council of China has put forward requirements for state-owned enterprises to speed up the construction of group data governance systems, build data governance systems and promote data governance work. It can be seen that data governance [2] has received great attention. The aerospace enterprise has the characteristics of small batch, refined and discrete [3]. With the development of aerospace manufacturing industry, the complexity of enterprise systems has increased. Since the number of enterprise application systems had increased, the amount of data has also shown explosive growth. In order to support the development needs of the aerospace enterprise, they have gradually carried out informatization construction. However, in the process of informatization construction, various data problems have become more and more obvious, such as scattered data, multiple sources, inconsistent data, numerous interfaces, and inconsistent standard rules.

Data resource directory is one of the most important step to build the integral enterprise data capability. It is also an important basis for subsequent data governance work,

B. Hu et al. (Eds.): BigData 2022, LNCS 13730, pp. 9–17, 2022.
https://doi.org/10.1007/978-3-031-23501-6_2

which is of great significance for data resource integration and improvement of data utilization [4]. Therefore, in order to solve the problem existing in the current data resource management process of the aerospace enterprise, it is necessary to carry out the sorting out of the data resource directory of the aerospace enterprise. Based on the analysis of the status quo of data resource management in the aerospace field, we study the data resource directory system of the aerospace enterprise from three aspects: data standard design, data resource inventory, and data resource directory framework construction. Finally, we introduce the application and practical result of the data resource directory system of the aerospace enterprise.

2 The Data Resource Directory System

The data resource directory system of the aerospace enterprise mainly includes data standard design, data resource inventory, and data resource directory framework construction, which is shown in Fig. 1.

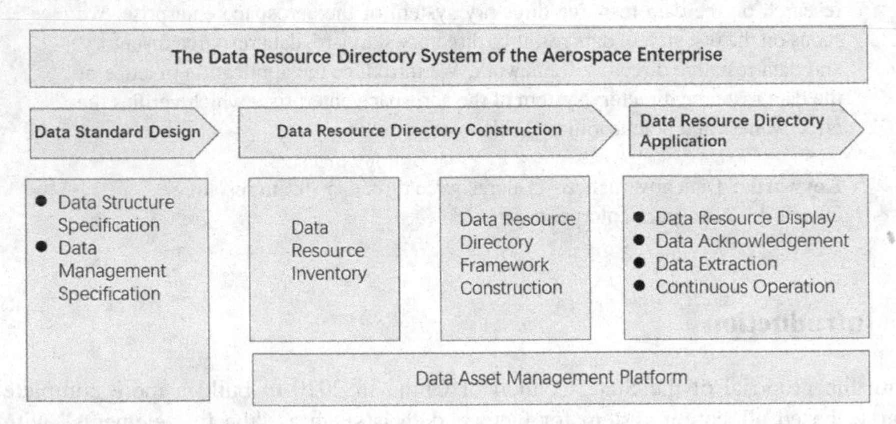

Fig. 1. The data resource directory system of the aerospace enterprise.

2.1 Data Standard Design

The construction of the data resource directory of the aerospace enterprise follows the principle of standards, which includes two parts: data structure specification and data management specification. The data structure specification studies the classification, coding and metadata of the data resource directory. The data management specification studies the maintenance process, organization and responsibilities of the enterprise's overall data resource directory.

Data Structure Specification

The data structure specification of the aerospace enterprise stipulates the hierarchical division and coding rules of the data resource directory.

The hierarchy division of the data resource directory of the aerospace enterprise refers to Huawei's data asset directory hierarchy (Appendix A) [5]. The hierarchy is divided into five layers from high to low: subject domain grouping, subject domain, business objects, logical data. Entities, attributes. The localization definition is formulated based on the characteristics of the aerospace enterprise management and engineering domain data resource.

(1) Subject domain grouping: The subject domain grouping of the aerospace enterprise is defined as the top-level data resource classification around the core business of FH production, reflecting the objective and logical collection of human, financial and material business data resource of the aerospace enterprise, with globality, independence, stability and other characteristics, such as enterprise foundation, business operation, comprehensive security, etc.
(2) Subject domain: The subject domain of the aerospace enterprise is defined as a data collection with similar internal business logic, strong data correlation, and high performance of internal departments and positions in management practice. It is the conceptual abstraction and data classification of specific businesses. High independence, basically non-overlapping, and highly overlapping management functions of business departments, such as human resource management, financial management, scientific research and production management, etc.
(3) Business object: The business object of the aerospace enterprise is defined as the data collections that carry the main business operation processes and activities in the business that is repeatedly executed by the business department and whose logic is relatively fixed, with obvious business logic correlation, technical process standards, and information. Data storage management and other characteristics, such as contract initiation, contract preparation, contract signing, etc.
(4) Logical data entity: The logical data entity of the aerospace enterprise is defined as a collection of data attributes with strong correlations generated in production management, which is an important carrier of the data resource and a core unit of management, such as contract negotiation minutes, contract approval forms, contract ledger, etc.
(5) Attribute: The attribute of the aerospace enterprise is defined as the lowest component element of the data resource, which describes the properties and characteristics of the data resource at the field level. It is the smallest particle of the data resource management and can be used as the basic element of the data resource. Data is stored, such as contract applicants, contract application departments, etc.

The data resource directory coding of the aerospace enterprise adopts 18-bit fixed length coding, which is managed in two levels, both of which use letters or numbers. The first level covers subject domain grouping, subject domain and business objects. The value range of subject domain grouping is "A0–Z9". The value range of subject domain is "AA–ZZ". The value range of business object is "000000–999999". The second level covers logical data entities and attributes. The value range of logical data entities is "AA00–ZZ99". The value range of attributes is "0000–9999".

Data Management Specification
The data management specification of the aerospace enterprise studies the management and control organization, responsibilities, maintenance process and other contents of the data resource directory system.

The management and control organization [6] includes three roles: data provider, data manager, and data demander. The roles and responsibilities of each category are designed as follows:

(1) The data providers are responsible for the planning and maintenance of the department's data resource directory. They are responsible for the collection, sorting and cataloguing of the department's data resource. They are responsible for collecting and updating the content of the data resource directory.
(2) The data management party is responsible for the overall planning of the data resource directory of the whole hospital. They are responsible for the directory review, registration, release, sealing and unsealing of the data resource directory content. They are responsible for the application, use and review of the data resource directory content. They are responsible for providing the query service of the data resource directory content.
(3) The data demanders uses the acquired information within the scope of authorization. They are responsible for setting changes to the content of the department's data resource directory.

The maintenance process of the data resource directory [7] covers the maintenance process of data addition, update and deletion. The specific process is shown in Fig. 2.

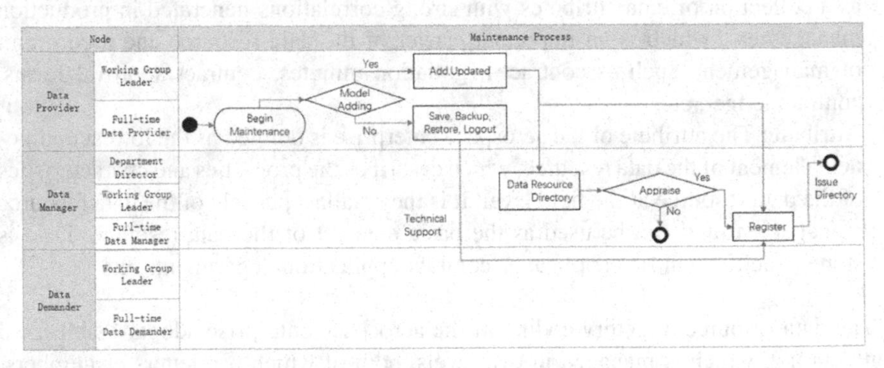

Fig. 2. The maintenance process of the data resource directory of the aerospace enterprise.

2.2 Data Resource Inventory

Data resource inventory is an important link in the sorting out of the data resource directory of the aerospace enterprise. Through the data resource inventory technology and the research of business personnel, the data structure, data type, storage method and other information in each heterogeneous system can be clearly grasped. The inventory process involves many aspects such as scope, organization and template design. We focus on studying the inventory template, combined with the characteristics of the aerospace enterprise business management and data resource, the data resource inventory template is shown in Table 1. The information of the data resource directory is presented in three dimensions: business, technology, and management.

Table 1. The data resource inventory template of the aerospace enterprise.

Number	Name	Classification	Required
1	Data object name	Business	Yes
2	Data resource description	Business	No
3	Field name	Business	Yes
4	Field source	Business	Yes
5	System contractor	Business	Yes
6	Data collection frequency	Business	Yes
7	Fata security	Business	Yes
8	Data collection range	Business	Yes
9	Fata focal point	Business	Yes
10	Network	Business	Yes
11	Shared level	Management	Yes
12	Sharing conditions	Management	No
13	Shared scope	Management	No
14	English table name	Technical	Yes
15	Field English abbreviation	Technical	Yes
16	Field type	Technical	Yes
17	Field length	Technical	Yes
18	Data storage format	Technical	Yes
19	Category	Technical	Yes

2.3 Data Resource Directory Framework Construction

Referring to the directory sorting methodology of mature industries and the standard design of the aerospace enterprise data resource, combined with the characteristics of the aerospace enterprise data resource and internal research, the data resource directory of the aerospace enterprise is sorted according to a five-level framework. The subject domain grouping includes strategic planning, business operations and Management support. The subject areas include planning, quality, scientific research. The business objects include asset operation management process, quality problem information management process, experiment management process. The logical data entities include contract statistics, contract negotiation minutes, etc. The attributes include contract name, applicants, etc. The schematic diagram of the data resource directory framework is shown in Fig. 3.

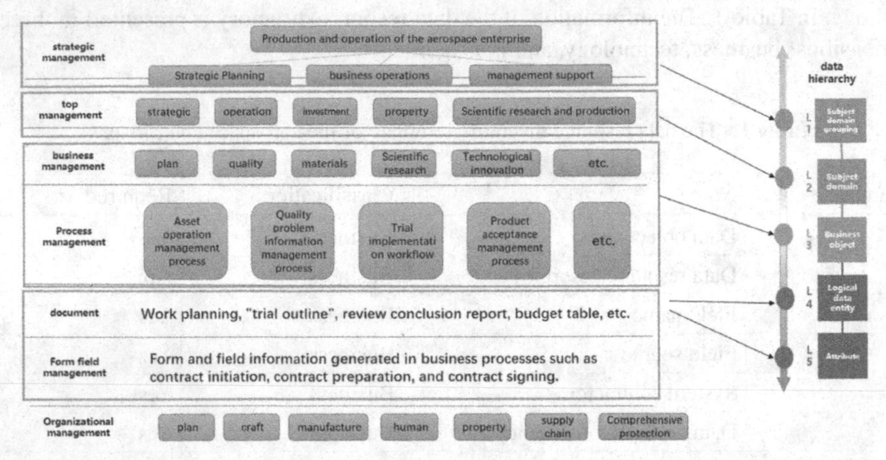

Fig. 3. Schematic diagram of the data resource directory framework of the aerospace enterprise.

3 Application

According to the design of the data resource directory system, select two representative fields (product BOM and contract) to carry out data resource inventory, sort out business processes and data sharing requirements, define the full life cycle of the data resource, and form data resource directory, import the directory into the data asset management platform. And the data information of the pilot business domain from the data warehouse, MES, contract and other business systems according to the directory information to form a data resource pool.

3.1 Data Asset Management Platform

The data asset management platform is the landing support tool for enterprise data resource management. The functions of the data asset management platform [8] built by the aerospace enterprise include metadata management, data standard management, data resource directory management, data services, data security and other functions. The platform manages the data resource directory online to better provide services to the outside world. The platform function architecture is shown in Fig. 4.

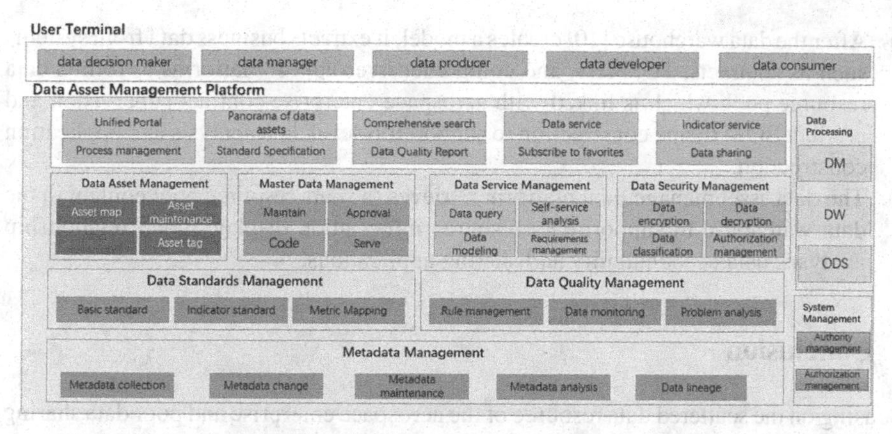

Fig. 4. The functional architecture of the data asset management platform.

3.2 Data Resource Pool Construction

Based on the data asset management platform, online directory management is carried out, and data is extracted from contracts, MES and other systems [9] according to directory information to form a data resource pool. The overall technical path is shown in Fig. 5.

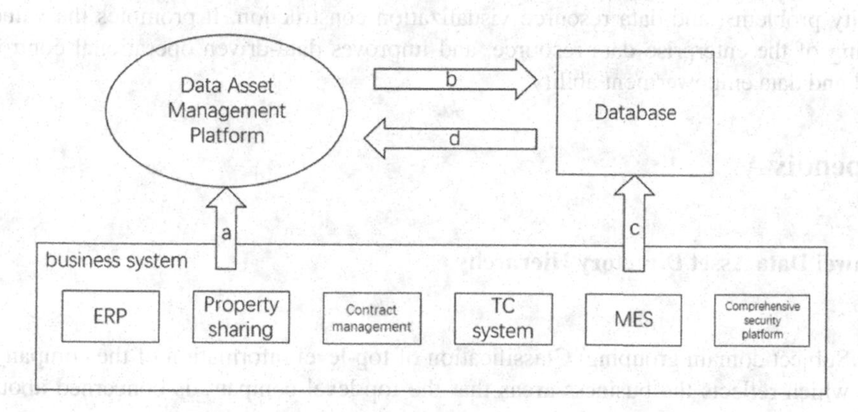

Fig. 5. The technical path of the data resource pool.

(a) The data asset management platform conducts model configuration according to the data resource directory structure specification and management specification, extracts the metadata information of ERP, financial sharing, contracts and other business systems, and automatically generates a data resource directory.
(b) After the data asset management platform generates the directory, the API service is released, and SAP BW calls the API service to obtain directory data and create a data model.

(c) After the data warehouse [10] creates a model, it extracts business data from systems such as contracts, TC, MES, and comprehensive support platforms to form a data resource pool, which is linked with aerospace enterprise contract supervision and product BOM theme construction to support top-level data analysis and application construction.
(d) The data asset management platform retrieves the data resource pool content in the data warehouse to support data resource information display, blood relationship analysis, data accountability and sharing applications.

4 Conclusion

Focusing on the scattered data resource of the aerospace enterprise and poor data sharing across enterprises and units, we study the data resource directory system of the aerospace enterprise from three aspects: data standard design, data resource inventory, and data resource directory framework construction. In according to the business need and the data resource of the aerospace enterprise, the data resource directory of the aerospace enterprise was built with five levels. And pilot fields were selected to carry out application practice through the data asset management platform. This verify the effectiveness and practicality of the data resource directory system of the aerospace enterprise. Through the online management of the pilot directory and the precipitation of the data resource, it provides basic data support for subsequent data lineage analysis, traceability of data quality problems, and data resource visualization construction. It promotes the value mining of the enterprise data resource, and improves data-driven operational control level and data empowerment ability.

Appendix A

Huawei Data Asset Directory Hierarchy

(1) Subject domain grouping: Classification of top-level information of the company, which reflects the business areas that the top-level company is concerned about through the data perspective.
(2) Subject Domain: A high-level classification of non-overlapping data used to manage its subordinate business objects.
(3) Business objects: Important people, things, and things in the business field, which carry important information related to business operation and management.
(4) Logical data entity: A combination of attributes with a certain logical relationship.
(5) Attribute: Describe the nature and characteristics of the business object to which it belongs, and reflect the minimum granularity of information management.

References

1. Zeng, Z., Wang, L.: The basic system of data factor market: outstanding problems and construction ideas. Macroecon. Res. (3), 17 (2021)
2. Correia, A., Gua, P.B.: A holistic perspective on data governance. In: Corporate Governance: A Search for Emerging Trends in the Pandemic Times (2021)
3. Liu, W.: Business-driven data governance method for aerospace enterprises. Inf. Technol. Informatization (5), 4 (2019)
4. Guohui, S.U., Dai, Q., Wei, H., et al.: Services of marine geology data resource directory based on REST and OData. Mar. Geol. Front. **34**(3), 26–32 (2018)
5. Huawei Data Management Department. Huawei's Way of Data. Machinery Industry Press (2020)
6. Lin, H.S.: Construction organization management and quality control of highway pavement construction. Constr. Des. Eng. (2018)
7. Nugroho, H., Gumilang, S.F.: Recommendations for improving data management process in government of bandung regency using COBIT 4.1 framework. In: ICSCA 2020: 2020 9th International Conference on Software and Computer Applications (2020)
8. Sun, Y., Jiang, Y., Li, B., et al.: Construction and application of rail transit security big data platform. Police Technol. (4), 4 (2020)
9. Herlyn, W.J.: The concept of the 'digital control twin'-impacts on the functioning mode and performance requirements of future ERP/MRP/PPS-systems structure of the presentation (2021)
10. Han, X.: Analysis on the Construction and Application of Data Warehouse in Mining Enterprises (2021)

Performance Analysis of Cross-Border Mergers and Acquisitions of Baiyuan Pants Industry in the Internet Era

Xingrui Yang[1] and Erna Qi[2(✉)]

[1] Huanggang Normal University, Huanggang 438000, Hubei, China
[2] Shaoguan University, Shaoguan 512005, Guangdong, China
drqimr@gmail.com

Abstract. The in-depth development of the mobile Internet has driven the innovation of enterprise value creation methods, and the cross-border integration of enterprises has gradually become an upsurge. This paper uses the financial index method to analyze the performance of Baiyuan Pants Industry's cross-border merger and acquisition of Global Tesco, and finds that it improves the performance as a whole after the merger. Then this paper analyzes the factors that affect the performance of cross-border M&A from the perspective of resource arrangement, and finds that Baiyuan Pants has broadened the development space by acquiring complementary assets, expanded existing capabilities through resource bundling and reorganization, and achieved leverage effect through resource deployment and utilization. After the cross-border M&A, Baiyuan Pants has reversed the continuous decline of the traditional clothing business through resource allocation, and realized the improvement of financial performance. The research in this paper can provide guidance for enterprises to achieve performance improvement through cross-border M&A, and can also provide a reference for the government to formulate industrial spatial layout and digital transformation policies.

Keywords: Internet era · Cross-border M&A · Performance analysis

1 Introduction

The Internet has changed the trading venue, expanded the trading time, enriched the trading varieties, accelerated the trading speed, and reduced the intermediate links [1]. Internet technology has enabled businesses to connect more globally than ever before, allowing organizations to unite across disciplines, geographies and industries. Compared with the PC Internet, the characteristics of the mobile Internet drive the new development of enterprise value creation methods. It completely realizes decentralization, and it is easy for enterprises to break the "pyramid" structure formed by powerful incumbent enterprises in the industry [2]. In the Internet era, "crossover" has become a hot issue, and some enterprises have even subverted the traditional industry operation mode after crossover [3]. Developed economies have exploded in the upsurge of cross-border integration of the Internet. And information technology enterprises and enterprises in other

B. Hu et al. (Eds.): BigData 2022, LNCS 13730, pp. 18–26, 2022.
https://doi.org/10.1007/978-3-031-23501-6_3

industries are competing for the forefront of industrial innovation around the new needs of cross-border integration, which has spawned a large number of new technologies, new formats and new growth points of the Internet economy, which are changing the development pattern of the whole industry and the global competition pattern.

In recent years, China's economy is facing transformation and upgrading. Due to inherent limitations in traditional industries, it is difficult to seek value breakthroughs within the industry. Therefore, it is expected to realize rapid transformation of enterprises through cross-border M&A of emerging industries, and absorb high-quality resources and advanced technologies to make enterprises back to life [4, 5]. Cross-border M&A have become a trend of corporate diversification strategies. In 2011, there were 577 cross-border M&As in China, and the growth rates in 2014 and 2015 were 75% and 59%, respectively. In 2019, Chinese M&A transactions fell to 2014 levels due to huge market uncertainty and the domestic deleveraging process limited financing channels (PwC, 2020). However, the significance of cross-border M&As to enterprise innovation, transformation and upgrading cannot be ignored. Especially, it is of great significance in promoting enterprises to break through the low-end lock-in of the global value chain, to achieve value chain climbing and cross-industry upgrading [6, 7].

So, for companies that continue to launch M&As across industry boundaries in the Internet era, can they improve performance? If the answer is yes, how do they improve performance for them? What is its process and mechanism? This paper intends to explore whether cross-border M&As of enterprises with asset-specific characteristics in the Internet age can improve performance through the case of Baiyuan Pants Industry cross-border M&A of Global Tesco. From the perspective of resource arrangement, this paper intend to deeply analyzes the factors affecting the performance of cross-border M&As and their mechanisms, thus it could provide guidance for enterprises to achieve performance improvement through cross-border M&As, and to provide a reference for the government to formulate industrial spatial layout and digital transformation policies.

2 Overview of the Case of Baiyuan Pants Industry' Cross-Border M&A of Global Tesco

Company profiles of both parties in cross-border M&As. Shanxi Baiyuan Pants Clothing Chain Wholesale Management Co., Ltd. (hereinafter referred to as "Baiyuan Pants Industry") was officially registered and established in Taiyuan City, Shanxi Province, China in 1995. Its main business is clothing retail business, including trousers and jeans, casual pants, etc., its target customer groups are middle-aged and young people in third- and fourth-tier cities in China. On December 8, 2011, Baiyuan Pants Industry was successfully listed on China's Shenzhen A-share small and medium-sized board (stock code 002640). Baiyuan Pants Industry is not only the first large-scale joint venture private enterprise in Taiyuan, Shanxi province, China, that has successfully listed new shares on the small and medium-sized board, it is also China's first large-scale private-owned professional knitting pants manufacturers listed through the listing of new shares. In 2014, Baiyuan Pants Industry cross-border M&A of Global Tesco started the transition to cross-border E-commerce development. Global Tesco was established in Shenzhen in

2007. The company mainly sells products, such as clothing, electronic products and various watches and toys, through its own cross-border E-commerce platform, its business spreads all over the world, and it ranks first among the major cross-border E-commerce companies in China.

2.1 Motivation for Cross-Border M&A of the Baiyuan Pants Industry

With the development of the economy, the prices of domestic raw materials and labor keep rising. However, the domestic traditional clothing manufacturing industry has extremely poor revenue capacity due to the lack of economies of scale advantages, single clothing styles, and lack of online sales channels, and even forced to close some offline stores. On the other hand, China's cross-border E-commerce platform is booming, and the import and export clothing products on the platform are favored by customers due to their high quality and low price, which promotes the continuous expansion of global online sales channels. At the same time, the state has continuously introduced favorable policies for the E-commerce industry in terms of macro policies, such as the Measures for the Administration of Online Transactions and the Opinions on Promoting the Innovative Development of Internet Finance. Considering the development difficulties of the pants industry and the broad prospects of cross-border E-commerce platforms, Baiyuan Pants industry decided to integrate offline domestic sales and cross-border trade through the cross-border M&A of Global Tesco, and take a bigger step towards transformation and upgrading.

2.2 The Process of Cross-Border M&A of the Baiyuan Pants Industry

On July 17, 2014, Baiyuan Pants Industry released the announcement of the draft M&A of Global Tesco; on September 19 of the same year, it released the announcement of obtaining the approval document of the China Securities Regulatory Commission; and on October 31 of the same year, the announcement of the completion of asset delivery was released. In this M&A, Baiyuan Pants Industry acquired 100% equity of Global Tesco for 1.032 billion yuan, with a premium rate of 796.98%; Baiyuan Pants Industry paid in both equity and cash; the founders of Global Tesco, Xu Jiadong and Li Pengzhen make performance commitments for 2014–2017, which are 65 million yuan, 91 million yuan, 1.26 billion yuan, and 1.27 billion yuan respectively. After the merger was completed, Baiyuan Pants Industry changed its name to "Global Top E-Commerce".

3 Performance Analysis of Cross-Border M&A of the Baiyuan Pants Industry

Because the financial performance of M&A investment has a certain lag, especially the cross-border M&A also involves the integration of special assets; therefore, this paper adopts the financial performance index method that can reflect the overall change nature and trend of the financial and operating conditions of enterprises. The performance of cross-industry M&A is analyzed. The data selection window is from 2013 (the previous year of cross-border M&A) to 2017 (the third year after cross-border M&A). The Financial Performance Index selects the enterprise profit, the debt service, the operation, the growth ability four aspects indicators. The main performance indicators of cross-border M&A in the Baiyuan Pants Industry are shown in Table 1.

Table 1. Main performance indicators of cross-border M&A of the Baiyuan Pants Industry

Indicator type	Indicator name	2013	2014	2015	2016	2017
Profit ability	ROE (%)	4.62	4.18	8.81	13.55	16.22
	Gross Profit Margin (%)	6.34	52.52	55.60	48.34	49.77
	Net Interest Rate (%)	6.41	3.40	4.20	5.01	5.47
Solvency	Flow Ratio	2.89	3.13	1.78	2.36	2.48
	Quick Ratio	2.37	2.33	1.02	1.27	1.07
	Assets and liabilities (%)	28.30	16.92	34.56	40.22	42.81
Operating Capacity	Accounts Receivable Turnover (times)	1.82	2.56	12.35	20.50	22.91
	Inventory Turnover Rate (times)	1.90	1.84	3.19	2.61	2.18
	Total Asset Turnover Rate (times)	0.48	0.53	1.49	1.67	1.79
Growth Ability	Operating Income Growth Rate (%)	-8.05	88.60	370.51	115.53	64.20
	Net Profit Growth Rate (%)	-39.86	6.22	403.39	133.85	90.72
	Growth Rate of Total Assets (%)	10.58	125.50	41.61	127.67	21.0

Data source: 2013–2017 annual report data of Baiyuan Pants Industry

As can be seen from the various financial indicators in Table 1, the cross-border M&A of Global Tesco by Baiyuan Pants Industry has achieved an overall improvement in performance. From the perspective of profitability, the return on equity and gross profit margin of Baiyuan Pants Industry has increased significantly compared with those before the cross-border merger, but the net profit rate has not been significantly improved. The reason may be related to the expenses incurred by the company to expand its business scale after the merger. From the perspective of solvency, the flow ratio and quick ratio of the Baiyuan Pants Industry have decreased compared with those before cross-border M&A, while the asset-liability ratio has increased; however, the corporate flow ratio can be stabilized at around 2, and the quick ratio has always been More than 1, the asset-liability ratio has never exceeded 50%, which indicating that the short-term and long-term solvency of the company is not bad. But the reason why the number of short-term debt-paying ability indexes of enterprises has decreased and the number of long-term debt-paying ability indexes has increased, the reason may be that after cross-border M&A, enterprises borrow from banks and other financial institutions in order to expand the scale of production and business scope, but reasonable liabilities can make enterprises play a more effective role in financial leverage. From the perspective of operational capabilities, the three indicators of the account receivable turnover rate, inventory turnover rate and total asset turnover rate of Baiyuan Pants Industry have shown an increasing trend compared with those before cross-border M&A, especially the accounts receivable turnover rate and the improvement of the two indicators of the total asset turnover ratio is particularly obvious, which shows that the problem of unsalable goods in inventory of the enterprise has been solved, the management efficiency of accounts receivable has been improved, and the operating capacity of the enterprise has been continuously strengthened.

From the perspective of growth ability, compared with before the cross-border M&A, the three indicators of the operating income growth rate, net profit growth rate and total asset growth rate of Baiyuan Pants Industry showed an overall growth trend, especially the growth rate of operating income and net profit. The improvement of the two indicators is particularly obvious; this shows that after the merger, Baiyuan Pants Industry has expanded its sales channels and business scope, expanded the company's scale, broke through the predicament of continuous decline in performance before the merger, and its growth competitiveness has also been significantly enhanced.

4 Analysis of the Reasons for the Successful Cross-Border M&A of the Baiyuan Pants Industry

4.1 Building a New Development Pattern by Acquiring Complementary Assets

Taking into account the rising raw material and labor costs of the apparel industry, the backward industry structure, the slow technological update, and the changes in the consumption patterns and consumption habits of potential customers in the Internet era, the Baiyuan Pants Industry feels that it is more and more difficult to capture to capture consumer preferences and dig their hidden needs in traditional ways. In order to get out of the dilemma, Baiyuan Pants Industry began to seek complementary assets, hoping to acquire online platforms or channels, and expand the development space through the integration of online and offline. Global Tesco is one of the largest cross-border E-commerce companies in China. It has a number of self-built vertical network sales platforms, and its sales network covers more than 20 countries. Its main products include casual clothing, electronic products, toys, etc. The company's website has more than 12 million registered users. After the acquisition is completed, Baiyuan Pants Industry can import its own products on the platform and online channels of Global Tesco, and use its big data capabilities to gain insight into consumer preferences to achieve product design innovation and precision marketing.In this way, Baiyuan Pants Industry has built a sales ecosystem of online and offline linkage, domestic and overseas synergy, which not only quickly reduces inventory pressure, but also greatly expands the market scope and development space. Global Tesco can take advantage of Baiyuan Pants Industry' supply channels and supply chain management to reduce clothing procurement costs and capital utilization efficiency, and use Baiyuan Pants Industry' offline stores to strengthen contact with consumers. In the second year after the completion of the merger, in order to speed up the layout in the E-commerce industry, Global Top E-Commerce (renamed as "Global Top E-Commerce" in 2016 by Baiyuan Pants Industry) successively participated in Qianhai Patoxun, Guangzhou Bailun and Tongtuo Technology, Cross-border Wing and other companies with complementary advantages, and then it gradually completed the asset acquisition of Qianhai Patuxun to strengthen its overall layout in the cross-border field.

4.2 Existing Capabilities Have Been Expanded Through the Bundling and Reorganization of Resources

After the merger and acquisition, Global Top E-Commerce will integrate all links in the company's business system, combine all business links with the advantageous resources such as logistics, payment, and promotion channels in the cross-border E-commerce business system, and actively explore new business cooperation models. Lay the foundation for the company's deep and horizontal layout in the industry. In terms of supply chain management, with the help of the rich supply chain management experience of Baiyuan Pants Industry, Global Top E-Commerce started to grasp the key links of quality control and integration of supply chain resources, and analyzed and applied big data throughout the company's sales, traffic import, procurement, logistics, customer service and other full business chains to achieve efficient supply chain control. In terms of stocking, Cross Border relies on the product market data and customer behavior analysis data provided by the database to find marketable products that meet the needs of users and maintain reasonable stocking to ensure the success rate of product development. In terms of advertising push, the company builds a multi-dimensional big data analysis model for customers, media, commodities, etc., through comprehensive analysis of sales data, conversion data, user behavior data, etc., to ensure that the right products are pushed to the right users through the right media to achieve the effect of precise delivery. The company's cross-border export E-commerce business traffic conversion rates from 2014 to 2016 were 1.45%, 1.55%, and 1.59%, respectively, which is leading the industry.

In terms of product import, Global Top E-Commerce makes full use of the company's scale advantage of cross-border export B2C, explores the best-selling products in the market, and quickly imports the market's best-selling products into the company's platform through data collection and analysis to further expand the company's sales. The company's platform of clothing and apparel, 3C electronics, security products, home gardening, sports and other products have maintained rapid growth. In 2016, the company achieved operating income of 8.54 billion yuan, a year-on-year increase of 115.53%. The company's inventory and slow-moving products management obtains operational data through various business links, develops matching algorithm models for different process links, and conducts real-time dynamic management of inventory and slow-moving products.

The company's product quality control, delivery speed, after-sales service and other process links have been further optimized to improve product quality, distribution efficiency and customer satisfaction, which directly promotes the increase in the repeat purchase rate of customers and the significant increase in the number of registered users of the platform.

By the end of 2016, the total number of registered users of the company's platforms exceeded 70.18 million, the average monthly active users exceeded 15 million, and the repeat purchase rate of the company's main platforms was 36.21%. In order to better grasp the opportunity of the rapid development of cross-border E-commerce and improve the all-round strategic layout of the E-commerce field, in August 2015, Global Top E-Commerce established a self-operated platform for cross-border E-commerce "Wuzhouhui" (www.wzhouhui.com), officially entered the field of cross-border import E-commerce, and successively launched PC websites and mobile apps relying on the self-operated platform of "Wuzhouhui"; Global Top E-Commerce acquired Youyi E-commerce and actively deployed the domestic market.

4.3 The Leverage Effect is Achieved Through the Deployment and Utilization of Resources

Through the application of IT technology, Global Top E-Commerce has established a full-process automated office system, and through independent algorithm research and development, it collects, organizes and precipitates all kinds of big data related to the company's business. The company carries on the depth conformity and the mining to each business link management data, has established the commodity operation system from the product development to the product on the shelf through each kind of algorithm optimization, the process optimization, thus realized the company each business link highly effective operation. For example, after-sales service system, warehousing logistics service system, supplier coordination management system, online precision marketing system, online sales activities operation system, website Operation System and user operation system. With an excellent management team, the company has built a number of differentiated, multi-level and multilingual professional vertical E-commerce platforms such as Sammydress and Gearbest. Among them, Sammydress is mainly positioned in clothing and apparel, while Gearbest is mainly based in emerging markets and is positioned in electronics products. In addition to its own brands, the company has also established good cooperative relationships with brand suppliers such as Xiaomi, Huawei, Haier, Lenovo, and etc., which has enriched the categories of export products and provided more choices for overseas consumers. According to a report released by Brandz The ranking list of "Top 50 Chinese Overseas Brands in 2018" shows that Gearbest, a wholly-owned subsidiary of the company's self-operated channel brand ranked 22nd, and Zuful ranked 34th. Global Top E-Commerce is the only cross-border E-commerce enterprises in China at the same time two own-brand shortlisted enterprises. In order to improve the efficiency of cross-border commodity distribution, improve service standards, and improve customer satisfaction, Global Top E-Commerce actively builds a global warehousing and logistics system and continuously expands overseas warehousing business. By the end of 2017, Global Top E-Commerce's warehousing area reached 280,000 square meters. Of which, there are 6 domestic warehouses with a storage area of 200,000 square meters. Through cooperation with third parties, the 54 overseas warehouses have been established in 17 countries, with a storage area of 80,000 square meters. The analysis of the reasons for the improvement of cross-border M&A performance in the Baiyuan Pants industry is shown in Fig. 1.

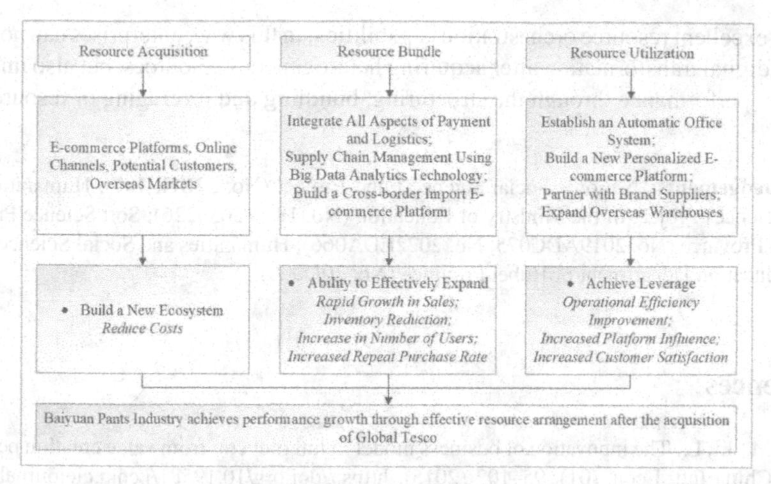

Fig. 1. Analysis of the reasons for the improvement of cross-border M&A performance in Baiyuan Pants Industry

5 Research Conclusions and Management Implications

Analysis Conclusion. This paper analyzes the case of Baiyuan Pants Industry' cross-border M&A of Global Tesco, and finds that the overall performance has been improved. Although traditional clothing business assets and E-commerce business assets are heterogeneous, they also have certain complementarities in the Internet era. For example, supply chain management experience and marketing channels are shared. If the two can effectively arrange resources, it can be build a business ecosystem with offline and online collaboration and domestic and overseas collaboration, thereby greatly improving operational efficiency and expanding market scope. The sample company Baiyuan Pants Industry in this paper, after acquiring the E-commerce company Global Tesco, has broadened its development space by acquiring complementary assets, expanded its existing capabilities through the bundling and reorganization of resources, and achieved leverage through the deployment and utilization of resources. And it has reversed the continuous decline of the traditional clothing business, and achieved the improvement of financial performance. After the merger, the number of registered users of the platform has also continued to increase, and customer satisfaction has gradually improved.

Management Inspiration. In the era of mobile Internet, the digital economy has maintained a strong growth trend, and almost two-thirds of unicorn companies are Internet companies or related to Internet businesses. The decentralized nature of the Internet provides a new way for enterprises to create value. If enterprises can rationally use big data and optimize algorithms, they can further improve their operational efficiency. From this perspective, Internet-related assets and traditional enterprises are highly complementary. Therefore, if the resources to be acquired by an enterprise are related to the Internet, before launching a cross-border M&A, the enterprise can estimate its own resource capacity gap in advance, analyze the number of its own contact points with the target resources, predict the possibility of complementary heterogeneous assets, and develop excellent resource orchestration capabilities; in this way, enterprises can not only achieve digital transformation after acquiring heterogeneous resources, but also improve company performance through the structuring, bundling and leveraging of resources.

Acknowledgements. National Social Science Fund Project (No. 22BGL107); Humanities and Social Sciences Project of the Ministry of Education (No. 18YJC630226); Soft Science Projects of Hubei Province (No. 2019ADC075; No. 2022EDA066); Humanities and Social Sciences Plan of the Education Department of Hubei Province (No. 20D098).

References

1. Luo, M., Li, L.: The innovation of business model in internet era: from value creation perspective. China Ind. Econ. (01), 95–107 (2015). https://doi.org/10.19581/j.cnki.ciejournal.2015.01.009.
2. Zhao, Z., Peng, H.: Crossover administration of "Internet Plus" — the theory building based on value creation. Sci. Res. Manag. **39**(09), 121–133 (2018). https://doi.org/10.19571/j.cnki.1000-2995.2018.09.014
3. Zhang, X., Wu, Q., Yu, X.: The logic of cross-boundary disrupt innovation of enterprises at internet age. China Ind. Econ. (03), 156–174 (2019). https://doi.org/10.19581/j.cnki.ciejournal.2019.03.019
4. Liu, P.: Research on cross-border mergers and acquisitions, social responsibility and corporate performance based on multiple cases. Mod. Bus. Trade Ind. **41**(04), 99–101 (2020). https://doi.org/10.19311/j.cnki.1672-3198.2020.04.049
5. He, X.: Cross - industry upgrading, strategy transformation and organization response. Stud. Sci. Sci. **37**(07), 1238–1248 (2019). https://doi.org/10.16192/j.cnki.1003-2053.2019.07.010
6. Yan, Y., Chi, R.: Re-study of the relationship between technical similarity and innovation performance after M&A—the moderating effect based on technology absorption capacity of enterprises. Sci. Res. Manag. **41**(09), 33–41 (2020). https://doi.org/10.19571/j.cnki.1000-2995.2020.09.004
7. Wang, W., Zhang, X.: Who is more likely to benefit from boundary - spanning technology M&A? Stud. Sci. Sci. **37**(05), 898–908 (2019). https://doi.org/10.16192/j.cnki.1003-2053.2019.05.016

Topic Modeling in the ENRON Dataset

Naciye Celebi(✉) and Narasimha Shashidhar

Department of Computer Science Sam Houston State University, Huntsville, TX, USA
{nxc038,nks001}@shsu.edu

Abstract. E-discovery is the electronic version of identifying, collecting, reviewing, and producing Electronically Stored Information (ESI) for the pre-trial procedure in a prosecution or legal investigation in many countries. There are challenges in e-discovery, such as cost and time consumption of the information retrieval process. Hence, natural language processing (NLP) is a key component in solving this problem, and we show using our case study that it scales effortlessly. Litigation costs are increasing, and as a result, legal professionals have sought to use fast information retrieval and machine learning methods to reduce manual labor and increase accuracy. In this paper, we consider using NLP to represent documents in a topic space using Latent Dirichlet Allocation and solving the information retrieval problem via finding document similarities in the topic space rather than doing it in the corpus vocabulary space. We also used the TF-IDF method in LDA to improve its performance. We report the results of our experiments on the ENRON dataset in the electronic discovery domain.

Keywords: E-Discovery · LDA · TF-IDF · Topic modeling · ENRON

1 Introduction

On a daily basis, legal departments of enterprises produce documents for documentation of cases. Whenever these enterprises are involved in litigation as part of regulatory requests, the need arises to obtain evidence (documents, tapes, etc.,) from the other party or parties. As the dataset of these documents increases, the enterprises turn to e-discovery technology to facilitate the process of collecting these documents. Electronic Discovery (e-discovery) is the disclosure of electronically stored information. Also, e-discovery is the electronic version of identifying, collecting, reviewing, and producing Electronically Stored Information (ESI) for the pre-trial procedure in a prosecution or legal investigation in many countries. Digitization of offices and the proliferation of smart devices have achieved rapid growth in ESI, such as emails, attachments, social media messages, etc. E-discovery process starts when a producing party is requested ESI for the legal departments of enterprises from information that it identifies as reasonably accessible. The request of obtaining evidence generally performed via a formal request for answers to interrogatories, request for production of documents (RPD), or request for admissions and depositions. A requesting party can

B. Hu et al. (Eds.): BigData 2022, LNCS 13730, pp. 27–34, 2022.
https://doi.org/10.1007/978-3-031-23501-6_4

obtain any information which relates to their need. The process of producing e-discovery steps started with production requests and followed by collection, review, and filtering process. The process concludes when the producing party provides all the responsive and not-privileged documents to the requesting party. Later, requesting parties conducts a manual analysis of all received documents. Earlier studies have shown that the majority of the cost in e-discovery is due to the manual review of ESI for responsiveness is 73% [12]. Sometimes even with expert information retrieval results of manual retrieval are inconsistent. Litigation costs are increasing, and as a result, legal professionals have sought to use fast information retrieval and machine learning methods to reduce manual labor and increase accuracy. In the e-discovery procedure, electronic documents identified potentially relevant by attorneys on both sides of a lawsuit are placed on legal hold - those documents than reviewed for relevance via a review platform after extraction and analysis. An accessible information retrieval approach for e-discovery is a keyword search or Boolean search. The first step of the process is to extract text within documents and data that describe documents. Second, each document, along with its extracted data fields, is indexed using an indexing engine. One popular alternative for this activity is the Octane Platform [1] index engine. The next task is to identify the best keywords to find relevant documents as quickly as possible. Documents retrieved from the indexing engine search will contain at least one of the search keywords. Generally, the indexing engine is an iterative method in which an attorney refines search terms repeatedly until they find all of the related documents. However, finding all of the relevant documents can be difficult and expensive. Therefore, both parties of a case that are using ESI must find a balance between the projected effort and potential benefits of the documents. Thus, in this study, we are interested in the document discovery or information retrieval part of the e-discovery process. We aim to provide Artificial Intelligence (AI) based technique to speed up the E-discovery information retrieval process and reduce its cost.

2 Background

The method of document retrieval is a comparison of incoming user queries with indexed text documents. A primary task is to represent entities, i.e., keyword queries and materials, in an indexing space where related entities lie near each other, and dissimilar ones are far apart. Salton et al. [2] proposed Vector space modeling, which is an indexing method that maps a corpus that consists of Documents and Vocabulary into a term-document matrix. Dumais et al. [3] proposed Latent semantic indexing (LSI) which is a method in NLP, in particular distributional semantics, of analyzing relations between a set of documents and the terms they contain by producing a set of concepts related to the documents and terms. Jones et al. [4] proposed term-frequency inverse document-frequency (TF-IDF), which performs over LSI as matrix factorization. Latent Semantic Indexing (PLSI) has applied to Enron corpus [11]. The authors examined the potential use of constructing social networks from email activity to generate insider threat leads. Blei et al. [5] proposed the Latent Dirichlet allocation (LDA) model, which is a probabilistic topic model. Berry et al. [8] proposed the Dolphin Search tool, which used for the documentation process of e-discovery. The tool splits the data into training and test sets. It receives a classification function information which maps the input documents to the similar labels using the training set. ENRON [10] dataset is suited for this method. Huang et al. [15] combined LDA with TF-IDF and came up with a new topic detection method named T-LDA. In their model, they selected keywords from microblogs contents according to TF-IDF weight. Secondly, they decide the appropriate K-value of the LDA and T-LDA on the basis of the Perplexity-K curve.

3 Methodology

In this paper, we consider representing ESI in a topic space using the well-known topic models such as Latent Dirichlet Allocation and solving the information retrieval problem via finding document similarities in the topic space rather than doing it in the corpus vocabulary space. We also used TF-IDF method along with Latent Dirichlet Allocation in order to improve its efficiency. Enron email set is used as a dataset in the experiment. The Enron email set is a large, publicly available dataset. The Federal Energy Regulatory Commission obtained it during its investigation of the Enron scandal. This dataset is a derivative of the FERC dataset and has been referenced in many email research studies and is also used by many commercial E-Discovery organizations. Therefore, ENRON dataset perfectly fits in our project. It is made of about 500.000 emails (1 GB of raw data) from 158 users, exchanged by the Enron's corporation employees from 1998 through 2002. Given its consequence, this corpus has been explored by a vast number of works from disciplines such as text mining and authorship attribution. This corpus is the right dataset for this study for different reasons. First, it is a realistic dataset about a real e-discovery case. Second, this dataset is large enough to perform LDA and TF-IDF models.

3.1 Text Preprocessing

The next step after data collection is to extract metadata by parsing ESI. For example, we can extract metadata such as from, to, subject, date, and email body from the ENRON email dataset. Metadata can give additional information for better indexing and efficient meta-field or faceted search. For any bag-of-words models such as TF-IDF or LDA, documents are required to be in plain text format and are converted into word tokens with a tokenizer. We use the python Natural Language Processing Toolkit (NLTK) [6] for tokenizing plain text. The performance of a text classification model is highly dependent on the words in a corpus and the features created from those words. Stopwords (common words) and other noisy elements increase feature dimensionality but do not help to differentiate between documents. There are some preprocessing steps that we used in our data to reduce the size of our corpus and add context prior to feature conversion. First step is removing stopwords; in natural language processing, useless words (data) are referred to as stop words. A stop word is a commonly used word (such as "some", "there", "which") that a search engine has been programmed to ignore, both when indexing entries for searching and when recovering them as the result of a search query [8]. We removed just generic stop words (such as "some", "there", "which") from our Enron dataset. Next step is lemmatizing words; lemmatization converts each word to its root form. Also, it returns the base or dictionary form of a word. (such as "player", "playing", "play") We lemmatized our data and received the root of each word.

3.2 Topic Modeling

Since our email dataset covered a wide range of topics, we utilized unsupervised learning to understand our data better. The topic modeling method allows us to describe and summarize the documents in a corpus without reading each article. It increases efficiency and saves time. It works by observing patterns in the co-occurrence of words using the frequency of words in each text. LDA assumes words are generated based off a Bayesian mixture of topic and word multinomial probabilities. Those multinomial probabilities are generated by Dirichlet distributions (with priors alpha and beta). By using an inference model (such as a Monte-Carlo Simulation), we can estimate the distribution of our model. This, in turn, allows us to draw some conclusions about documents and topics. Additionally, LDA assumes that these mixtures follow a Dirichlet probability distribution. This definition means that for each document, we can assume there should only be a few topics covered and that for each topic, only some words are associated with that topic. For example, in an article about social life, we would not expect to find many different topics covered. LDA model used to reveal topics under the assumption that it "generated" them–hence, we "find" the parameters of the model, we reveal the topic distributions. The model initializes randomly and updates topics and words as it repeats through every article to find a specific number of topics and associated corresponding words. The hyper-parameters alpha and beta can be modified to control the topic distribution in every article and word distribution per topic, individually. A high alpha means that every document is likely to contain a mixture of most topics and a high beta means that each topic is likely to contain a mixture of most words. LDA is entirely unsupervised, but the user must provide the model with a specific number of topics to describe the entire set of documents. To visualize our model, we chose 20 topics to see variety of topics. Our coherence score actual gives us a good clue to choose our number of topics. We want this score to be as high as possible for the greatest number of topics. Therefore, we selected 20 topics. Topics are not named, but we can get a better understanding of each topic by looking at the words associated with them.

3.3 Term Frequency-Inverse Document Frequency (TF-IDF)

Term Frequency-Inverse Document Frequency (TF-IDF) is a weighting factor used in data and text mining. TF-IDF is a numerical statistic that is intended to reflect how important a word is in a document or corpus. Term-Frequency (TF), the number of frequency of occurrences of a word in a document. If a word appears in the document many times that reflects the words are more important words. Inverse-Document-Frequency (IDF) is a measure of the word "weight". Based on the word frequency, if a word is less frequent in documents, it means that it is nearly rare; on the other hand, it appears many times in the document. The larger the IDF value of the word, the larger its weights in the document. Therefore, when a word is more common, their IDF is lower. The TF-IDF is reached by multiplying the two values. For example,"investigator" could be the key word because it does not appear in standard documents. But it becomes a common word in Digital Forensics in which it is mentioned frequently. A high weight in TF-IDF is reached by a high term frequency and a low document frequency of the term in all of the collection of documents. Hence, the weights tend to filter out standard terms.

4 Experiment Result

In this paper, we perform an experiment by applying TF-IDF to the ENRON dataset, followed by LDA, to get the best results in a variety of topics. Lastly, to distinguish different topics using the words in each topic and their corresponding weights, we run LDA with the TF-IDF method. Then, we report the performance of the results. In our ENRON dataset, we have roughly 500,000 emails tagged to 158 people. In this huge email dataset, we used the cleaned version of ENRON dataset which provided by Kaggle [7]. Emails in the dataset separated into 3 columns: index, messageid, and the raw message. Before working with this data, we parsed the raw message into key-value pair. Therefore, it helps us to work on the desired portion of the email, such as the sender, receiver, and email body data. Then, we extract the 'from', 'to', and the email 'body' from all the email content. With the usage of map-to-list method, we run through all the emails and combine them together into a single record. Then, we completed text preprocessing steps to make our dataset ready for TF-IDF method.

4.1 Analyzing ENRON Dataset with TF-IDF

TF-IDF is a numerical statistic that intended to reflect how important a word is into a document in a collection or corpus. TF is a term that frequently occurs in many documents, and IDF terms will be the ones that are used in a small set of documents. In our case, a "document" is all the emails in a folder. Thus, the numerator is the total number of folders, and the denominator is the number of folders where some email in the folder contains the term. Table 1 illustrates the top terms out of all the emails.

4.2 Analyzing ENRON Dataset with LDA

LDA assumes documents are created one word at a time. For each word, a probability distribution of distributions (Dirichlet) is sampled. Then that distribution is sampled to select a word. Many LDA implementations use monte

Table 1. Top terms in ENRON dataset

Features & Score	
enron	0.044036
com	0.033229
etc.	0.027058
hou	0.017350
message	0.016722
original	0.014824
phillip	0.012118
image	0.009894
gas	0.009022
jhon	0.008551

carlo methods to discover those distributions. A document is a mixture of diverse topics which itself is an expression of words all tagged with a given probability of occurrence. We have topic representations across all the documents and word distribution across all the topics. The main part of LDA is this concept of a Dirichlet distribution, and much like a Normal distribution, it is a probability distribution. One main difference is that in this case, instead of the probabilities being sampled over a space of real numbers, it is sampled over a probability simplex. At the same time, probabilities generated by LDA for the words is that a word can be reflected in multiple topics. Lastly, we used Matplotlib [13] and PyLDAvis [14] to visualize the output of our analysis. In Fig. 2, we visualize the output. Here are some observations; the size of the bubbles tells us how dominant a topic is across all the documents also, the words on the right-hand side are the keywords driving that topic and the closer and overlapping the bubbles, the more similar the topic. The farther they are apart, the less similar (Fig. 1).

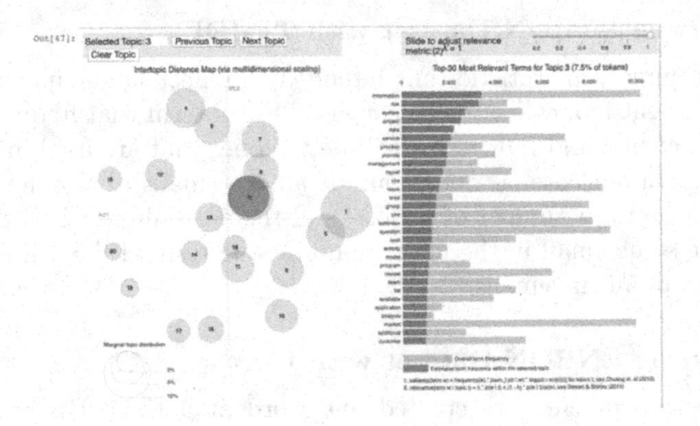

Fig. 1. LDA model

As TF-IDF has a high calculate speed and does not take into account the positions of words in documents, we combined LDA with TF-IDF and applied on the ENRON dataset. TF-IDF generally used in Vector Space Model (VSM) for determining the weight of words in an article. It considers the word frequency and the number of the document that includes the words, which makes it helpful to select keywords in the ENRON dataset. The higher weight of the word, the greater ability it has for representing the basic idea of the document in the ENRON. In our LDA with the TF-IDF model, we used the TF-IDF method to calculate the weight of words in the preprocessed text. Therefore, the size of the word corpus can be reduced efficiently, and the keywords that indicate topic contents can be kept to the maximum amount. After the process of word selection, LDA with the TF-IDF model conducts the Monte Carlo search to produce the topic result. As for the combination of TF-IDF and LDA, we calculated the TF-IDF weight of each word before the Monte Carlo method and selected keywords according to the weight. Only the former N/10 words with higher weight were reserved, where N is the size of word sets. In order to verify the effectiveness of LDA in the aspect of capturing topics in the ENRON dataset, we analyzed the results by various evaluation indexes. Runtime records the time taken for the modeling process. The Precision indicates the ratio of results detected correctly in modeling results. Recall rate represents the ratio of correct results in the text. F-Measure is a comprehensive evaluation that combines Precision and Recall rate. The results are meaningful when the F-measure is high. It is apparent to find that TF-IDF and LDA run faster than LDA, which demonstrates the improvement of the size of the word sets is effective. The TF-IDF with LDA method faster as it analyzed the ENRON email dataset with a runtime of 35.88 s. On the other hand, LDA requires more time. The runtime of the LDA is about 201 s.

Table 2. Accuracy comparison of the ENRON dataset.

Method	Recall	Precision	F-score
LDA	0.45	0.46	0.65
LDA with TF-IDF	0.47	0.51	0.74

We used three different performed evaluation metrics: Recall, Precision, and F measure. Since we are dealing with multiple topic setup, these metrics were calculated based on Sokolova et al. [9]. Table 2 describes the performance of the three adopted measures. LDA has comparable F-score to the conceptual approaches with 0.65 F-score. However, its recall is 0.45, while its precision is0.46. The accuracy of the LDA with TF-IDF method we received the recall is 0.47, precision is 0.51, and F-score is 0.74.

5 Conclusion

Nowadays, AI-based automation system is essential in e-discovery. Likely, AI and Machine Learning become more widely used in the near future as parties gain reliance on its accuracy, and as we show some preliminary evidence that it truly speeds up the information retrieval process and reduces costs. It is probable to observe that topic modeling is needed in the e-discovery field. It was able to build clusters of similar emails without any human intrusions. In this study, we compared TF-IDF, LDA and LDA with TF-IDF approaches to topic detection using the Enron dataset. The results showed that LDA with TF-IDF outperformed the approaches, especially in terms of precision.

References

1. https://www.capterra.com/p/119799/Octane-Platform/
2. Salton, G., Wong, A., Yang, C.S.: A vector space model for automatic indexing. Commun. ACM **18**(11), 613–620 (1975)
3. Dumais, S., Furnas, G., Landauer, T., Deerwester, S., et al.: Latent semantic indexing. In: Proceedings of the Text Retrieval Conference (1995)
4. Jones, K.S.: A statistical interpretation of term specificity and its application in retrieval. J. Doc. **28**, 11–21 (1972)
5. Blei, D.M., Ng, A.Y., Jordan, M.I.: Latent Dirichlet allocation. J. Mach. Learn. Res. **3**, 993–1022 (2003)
6. Bird, S., Klein, E., Loper, E.: Natural Language Processing with Python. O'Reilly Media, Sebastopol (2009)
7. Cukierski, W.: The Enron Email Dataset. Kaggle, 16 June 2016. https://www.kaggle.com/wcukierski/enron-email-dataset/version/2#emails.csv
8. Berry, M.W., Esau, R., Keifer, B.: The Use of Text Mining Techniques in Electronic Discovery for Legal Matters, chap. 8, 174–190. IGI Global (2012)
9. Sokolova, M., Lapalme, G.: A systematic analysis of performance measures for classification tasks. Inf. Process. Manage. **45**, 427–437 (2009)
10. Irimia, R., Gottschling, M.: Taxonomic revision of Rochefortia Sw. (Ehretiaceae, Boraginales). Biodiv. Data J. **4**, e7720 (2016). https://doi.org/10.3897/BDJ.4.e7720
11. Okolica, J.S., Peterson, G.L., Mills, R.F.: Using PLSI-U to detect insider threats by datamining e-mail. Int. J. Secur. Netw. **3**(2), 114 (2008)
12. https://www.abajournal.com/advertising/article/reducing-costs-with-advance-review-strategies
13. Matplotlib.org. 2020. Matplotlib: Python Plotting - Matplotlib 3.2.1 Documentation. https://matplotlib.org/. Accessed 18 Apr 2020
14. PyPI 2020. Pyldavis. https://pypi.org/project/pyLDAvis/. Accessed 18 Apr 2020
15. Huang, L., Ma, J., Chen, C.: Topic detection from microblogs using T-LDA and perplexity. In: 2017 24th Asia-Pacific Software Engineering Conference Workshops (APSECW). IEEE (2017)

Design of Multi-data Sources Based Forest Fire Monitoring and Early Warning System

Xiaohu Fan[1,2](✉), Xuejiao Pang[1], and Hao Feng[1]

[1] Department of Information Engineering, Wuhan College, Wuhan 430212, China
{9420,9452,8206}@whxy.edu.cn

[2] Wuhan Bohu Science & Technology Co., Ltd., Wuhan, China

Abstract. The cause of forest fire is complex, which depends on meteorological, surface soil and human integrated monitoring network and multi-source data like remote sensing. This paper provides 4 kinds of monitoring method based on satellite, IoT, UAV aviation cruise, etc. Carrying out all-round fire point detection, combined with administrative division data, forest-grassland resource data and meteorological observation data, the fire monitoring and customized analysis are carried out through computer automation, computer vision technology and deep learning algorithms, the fire thematic functions are generated. It also supports the release and display of fire information in various forms such as SMS, email, Web terminal and mobile terminal, so as to grasp the occurrence of fire points in the area at the first time, and realize the 24-h uninterrupted forest fire monitoring and early warning. Multi-channel radiation fusion, yolov4 algorithm is used for training, combined with the IoT data, the accuracy of early warning and pre-assessment improved, provides scientific decision-making basis for forest fire prevention and rescue work.

Keywords: Fire point detection · Remote sensing · IOT · YoloV4

1 Introduction

According to the official data of the National Bureau of statistics of China, from 2015 to 2020, there were many fires, with a large area affected and a large number of people and direct losses [1]. In 2021, with the state's increasing attention to forest fire prevention and extinguishing, and the promotion and application of integrated space-based remote sensing, the number of fires and damage were significantly reduced, and no major fires occurred throughout the year.

On March 19, 2021, the State Forestry and grassland Administration held a national spring video and telephone conference on forest and grassland fire prevention, requiring forest and grass departments at all levels to thoroughly implement the decisions and arrangements of the Party Central Committee and the State Council, adhere to the concept of people first and safety first in accordance with the unified command of the National Forest Defense Office, and take the prevention of major casualties and property losses, and the occurrence of major forest and grassland fires as the goal, To realize the whole chain management of 'preventing failure, danger and violation'.

B. Hu et al. (Eds.): BigData 2022, LNCS 13730, pp. 35–51, 2022.
https://doi.org/10.1007/978-3-031-23501-6_5

On August 18, 2021, the State Forestry and Grassland Administration issued the 'fourteenth five-year plan for forestry and grassland protection and development -- joint construction of an integrated forest and grassland fire prevention and extinguishing system', which requires that in terms of improving the prevention system, improving the early warning ability requires comprehensive use of various monitoring means of 'sky to ground' to improve the ability to actively grasp the fire.

Therefore, based on the actual needs of forest and grass fire prevention, the new generation of information technologies such as satellite remote sensing technology, cloud computing, big data and mobile Internet are comprehensively utilized. Based on satellite remote sensing image data, ground video monitoring data and ground patrol report data, the integrated forest fire monitoring service of 'diversified data fusion' is constructed, which is conducive to early detection of fire and rapid disposal of fire.

2 Related Works

Barrile et al. fully combined GIS system technology to conduct post disaster analysis and judgment [1], Rahman et al. studied the model of GIS combined with remote sensing data and applied it in the northwest tribal land of Bangladesh [2]. Bourjila et al. carried out comprehensive assessment of cases in Morocco using GIS, remote sensing data and groundwater data [3]; Chen et al. used multi GBDT infrared thermal channels to monitor forest fire points in Yunnan, China [4], and Kumar et al. also used similar remote sensing and GIS systems to realize forest fire monitoring applications in Guntur area [5]. Boselli used remote sensing to monitor and study the aerosol characteristics of the mega fire in Vesuvius, Italy [6]. With regard to the IoT, Kanakaraja uses the ubidot platform [7], Singh H further adds the concept of cloud side collaboration [8], Singh R proposes the concept of Digital Forest 4.0 [9], Sun and others further strengthen the monitoring means in combination with UAV and MEC [10, 11]. Gaitan et al. proposed the IOT method of Lora networking to realize the smart forest [12].

To sum up, the previous research methods are generally to improve the data dimension and the monitoring coverage area, fully reduce the manpower and improve the response speed of fire warning, so as to minimize the expansion of losses.

3 Top-Level Design

See Fig. 1.

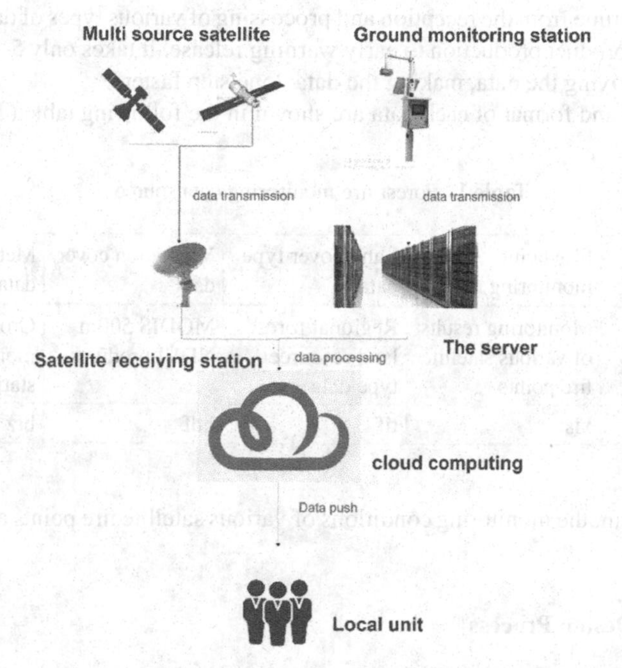

Fig. 1. Overall architecture of monitoring and early warning application system

Acquiring multiple types of data and combining with precise model algorithms, the specific data of forest fire points can be calculated. Once the threshold value is reached, the local unit can be notified in the first time to carry out timely early warning and rapid disposal, so as to reduce or avoid the loss of personnel and property caused by fire.

3.1 Data Source

Comprehensive monitoring means are adopted for high integration, satellite remote sensing, ground monitoring and other data are combined, so as to achieve four-in-one monitoring, create an all-round and all-time monitoring network without dead angle, process massive multi-source data, and automate the whole process to achieve high timeliness fire monitoring.

Integrate more than 10 kinds of satellite remote sensing, such as H8, FY, NPP, NOAA and MODIS, and video images taken by UAVs and observation tower cameras, analyze and identify, accurately capture fire points, so that the technology 'does not fight' and the fire points 'Nowhere to hide'. The process is completely automatic and the task is adjusted at any time from the reception and processing of various types of data, fire point identification, product production to early warning release. It takes only 5–10 min from the date of receiving the data, making the data "one step faster".

The source and format of each data are shown in the following table (Table 1):

Table 1. Forest fire monitoring data source

	Fire point monitoring results	Land cover type data	Vegetation cover data	Meteorological data
Data source	Monitoring results of various satellite fire points	Regional forest land resource type data	MODIS 500m NDVI products	Ground monitoring station data
Provide format	xls	tif	hdf	brz

Among them, the monitoring conditions of various satellite fire points are as follows (Table 2).

3.2 System Design Process

See Fig. 2.

3.3 Functional Design

Automatic Identification of Suspected Fire Points. By using the integrated forest fire monitoring means of space, space and earth, the whole process automation from satellite data reception and processing, high-precision fire point identification and information extraction to early warning information push is realized, and the 24-h real-time independent monitoring capability of forest fire based on satellite remote sensing and other technologies is formed. Among them, under normal conditions, the monitoring frequency of satellite whole area monitoring can be up to 5 min/time, and the minimum open fire area of the fire point that can be monitored can reach 10 square meters; Ground camera PTZ control, 360° real-time monitoring of ground conditions; The UAV can be used for forestry patrol and fire scene tracking at any time. The resolution of aerial images can reach 0.03 m.

Push Alert Information. When a suspected fire is found, it will automatically push the information of the suspected fire point to the business leader of the fire prevention and extinguishing department and the ground patrol personnel in the first time through SMS, e-mail, website and mobile terminal, so as to ensure that the business leader of the fire prevention and extinguishing department and the ground patrol personnel can receive the early warning information and make decisions in time.

Table 2. Satellite data source

	Sunflower No.8	Fengyun No.4	Gaofen No.4	TERRA	FengYun-3 C	Fengyun 3 B	FengYun-3 D	NPP	NOAA20	AQUA
Satellite type	Geostationary satellite	Geostationary satellite	Geostationary satellite	Polar orbiting satellite	Polar orbiting satellite	Polar orbiting satellite	Polar orbiting satellite	Polar orbiting satellite	Polar orbiting satellite	Polar orbiting satellite
Monitoring frequency	1 time/10 min	1 time/5–15 min	1 time/10 min	2 times/day	2 times/day	2 times/day	2 times/day	2 times/day	2 times/day	2 times/day

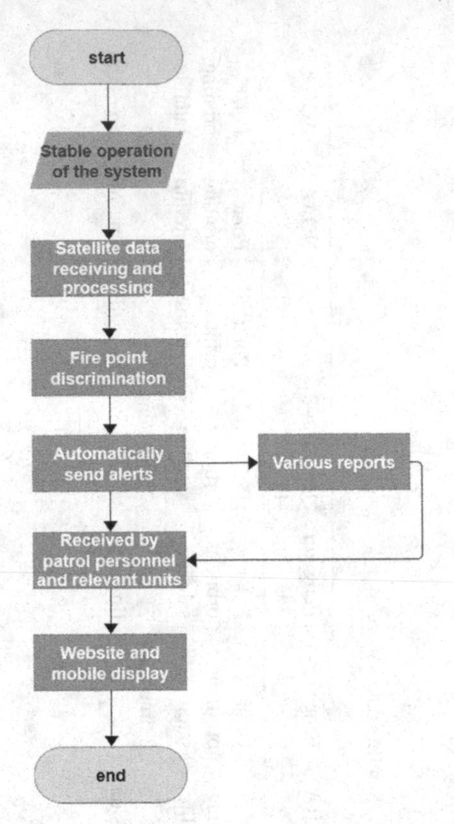

Fig. 2. Flow chart of forest fire monitoring service

Report by Ground Inspector. The ground patrol personnel shall carry out the ground inspection in a timely manner according to the early warning information, and then report the inspection results through the app for the convenience of the business principal.

4 Algorithm Design

4.1 Forest Fire Point Satellite Monitoring

The emissivity of the pixel observed by the satellite is the weighted average of the emissivity of all parts of the ground objects within the pixel range, namely:

$$I_t = (\sum_{i=1}^{n} \Delta S_i I_{Ti})/S \tag{1}$$

where I_t is the emissivity of the pixel observed by the satellite, t is the brightness temperature corresponding to the emissivity N_t, S_i is the area of the ith sub region in the pixel, I_{Ti} is the emissivity of the sub region, T_i is the temperature of the sub region, and S is the total area of the pixel.

When there is a fire point on the ground, the emissivity of the pixel containing the fire point (hereinafter referred to as the mixed pixel) can be expressed by the following formula:

$$\begin{aligned} I_{imix} &= P * I_{ihi} + (1 - P) * I_{ihi} \\ &= P * \frac{C1V_i^3}{e^{C2V_i/T_{hi}} - 1} + (1 - P) * \frac{C1V_i^3}{e^{C2V_i/T_{bg}} - 1} \end{aligned} \tag{2}$$

where C1 = 1.1910659 × 10^{-5} MW/(m^2.Sr.cm^{-4}), C2 = 1.438833 K/cm^{-1}, where I_{imix} is the emissivity of mixed pixel, P is the percentage of the area of sub-pixel fire point (i.e. open fire area) in the area of pixel, I_{ihi} is the emissivity of fire point, I_{ibg} is the background emissivity around the fire point, T_{hi} is the temperature of fire point, T_{bg} is the background temperature, and I is the serial number of infrared channel.

According to Wien's displacement law:

$$T * \lambda \max = 2897.8(k, \mu m) \tag{3}$$

The temperature T is inversely proportional to the radiation peak wavelength, that is, the higher the temperature, the smaller the radiation peak. At normal temperature about 300K, the peak wavelength of surface radiation is 10.50–12.50 μ The combustion temperature of herbaceous plants and trees is generally above 550K, and the flame temperature is more than 1000K. The peak wavelength of thermal radiation is close to 3.5–3.9 μ M U.M wavelength range. Therefore, when there is a fire on the ground, the count value, emissivity and brightness temperature of the mid infrared band will change sharply, forming an obvious contrast with the surrounding pixels and far exceeding the increment of the far infrared channel. This feature can be used to detect the ground fire point of forest fire.

The figure below shows the mid infrared (3.7 λ M U.M) and far infrared (11 λ M U.M) as a function of temperature. It can be seen from the figure that when the temperature changes from 300K to 800K, the radiation of the mid infrared channel increases by about 2000 times, while that of the far infrared channel only increases by more than ten times (Fig. 3).

According to the daily fire monitoring experience and the results of the satellite ground Synchronization Experiment of the artificial fire site, when the mid infrared channel is 8K higher than the background bright temperature and the difference between the mid infrared and far infrared bright temperatures is more than 8K higher than the difference between the mid infrared and far infrared bright temperatures of the background, it is generally an abnormal high temperature point caused by open fire. The results of the satellite earth synchronous observation experiment at the artificial fire site in Wuming, Guangxi, show that the open fire area with an area of more than 100 m^2 can cause a temperature increase of about 9K in the mid infrared channel, reaching the identification threshold of daily fire monitoring. Therefore, the fire point conditions are mainly determined according to the brightness temperature increment of the mid infrared channel and the brightness temperature difference increment between the mid infrared channel and the far infrared channel. Take HJ-1B mid infrared channel ch7 and far infrared channel ch8 as examples:

$$T_7 - T_{7bg} > T_{7TH}, T_{78} - T_{78bg} > T_{78TH} \tag{4}$$

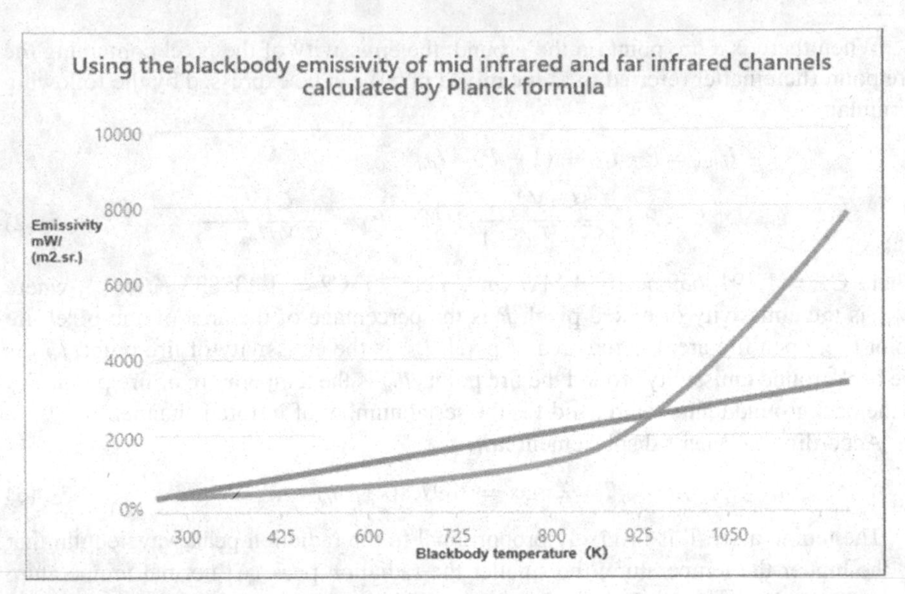

Fig. 3. Temperature dependence curves of blackbody emissivity in mid infrared and far infrared channels calculated by Planck formula

T_7, T_{7bg}, T_{7TH} are the mid infrared channel brightness temperature, the mid infrared channel background brightness temperature, and the mid infrared channel fire point identification threshold of the identified pixel. T_{78}, T_{78bg}, T_{78TH} are the difference between the mid infrared and far infrared brightness temperature of the identified pixel, the difference between the mid infrared and far infrared brightness temperature and the background brightness temperature, the difference between the mid infrared and far infrared brightness temperature, and the discrimination threshold of different fire points.

The background temperature calculation has a direct impact on the discrimination accuracy. For the dense vegetation covered area with single underlying surface, the average of adjacent pixels is representative of the identified pixels. However, in the ecotone between vegetation and desert, the average brightness temperature of adjacent pixels calculated from the vegetation coverage of each pixel may be significantly different from that of the identified pixel, so the identification threshold needs to be adjusted accordingly. Kaufman proposed a method to determine the discrimination threshold by using the standard deviation of brightness temperature of background pixels, namely:

$$T_7 > T_{7bg} + 4\delta T_{3bg}, T_{7bg} > \Delta T_{78} + 4\delta T_{78bg} \tag{5}$$

where T_7 is the brightness temperature of the mid infrared channel of the identified pixel, and ΔT_{78} is the difference between the brightness temperature of the mid infrared and far infrared channels of the identified pixel. T_{7bg} is the brightness temperature of the background mid infrared channel, and the T_{7bg} is brightness temperature difference of the background pixel between the mid infrared and far infrared channels, all taken from the average value of the surrounding 7 * 7 pixels. δT_{7bg} is the standard deviation of the brightness temperature of the infrared channel in the background pixel, and δT_{78bg}

is the standard deviation of the brightness temperature difference between the infrared channel and the far infrared channel in the background pixel, namely:

$$\delta T_{7bg} = \sqrt{\left(\sum_{i=1}^{n} (T_{7i} - T_{7bg})^2\right)/n} \tag{6}$$

$$\delta T_{78bg} = \sqrt{\left(\sum_{i=1}^{n} (T_{7i} - T_{8i} - T_{78bg})^2\right)/n} \tag{7}$$

where T_7 and T_{8i} are the brightness temperature of the ith mid infrared channel and far infrared channel of the peripheral pixel used for calculating the background temperature, respectively. When δT_{7bg} or δT_{78bg} is less than 2K, set it to 2K.

In the calculation of background brightness temperature, it is also necessary to remove the influence of cloud area, water body and suspected fire point pixels, that is, before calculating the average temperature, the cloud area, water body and suspected fire point pixels in the neighborhood are excluded, and only the pixels under clear sky conditions are used for calculation.

The judging conditions for the pixel of suspected fire point are:

$$T_7 > T_{7av} + 8K, T_{78} > T_{78av} + 8K. \tag{8}$$

T_{7av}: the average value of channel 7 with brightness temperature less than 315K after excluding cloud areas and water pixels in the neighborhood. T_{78av}: The average value of the brightness temperature difference between channel 7 and channel 8 when the brightness temperature of channel 7 is less than 315K after excluding cloud areas and water pixels in the neighborhood.

After removing the cloud area, water body and high temperature pixels, if there are too few remaining pixels (such as less than 4 pixels), the size of the neighborhood window can be expanded, as shown in 9*9, 11*11 eal.

Solar flares have a serious impact on the mid infrared channel. Generally, the pixels in the solar flare area are not used for fire point identification.

4.2 Forest Fire Point Video Monitoring

The target detection algorithm based on the deep learning model yolov4 is used to identify the fire point in each frame of the video, detect the fire point quickly and efficiently, and realize the annotation of the forest fire point location. The fire point identification under the infrared channel is realized by using the temperature represented by the image brightness as the main core basis and combining the morphological analysis of the flame shape. The specific process is as follows (Fig. 4):

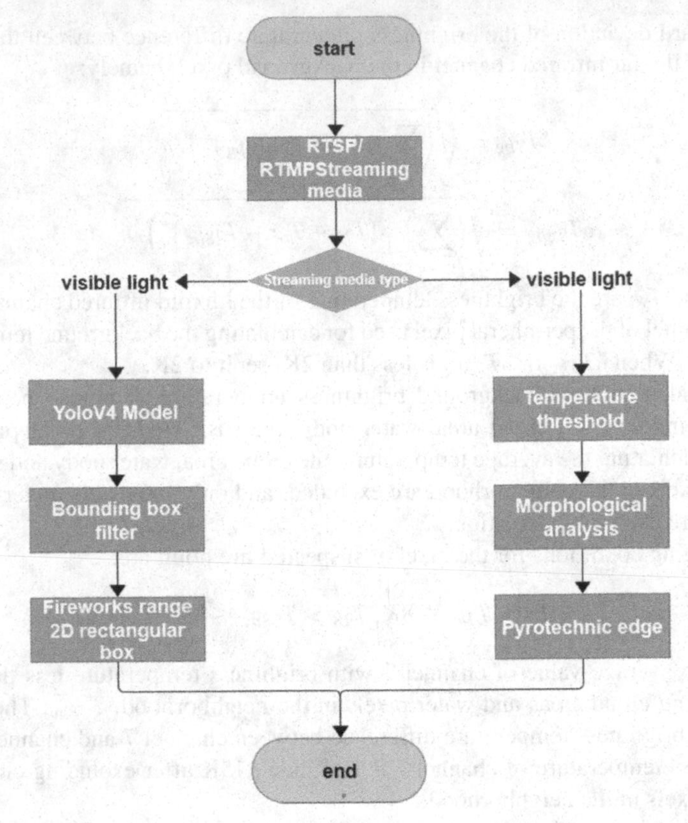

Fig. 4. Video fire point identification algorithm flow

Fire Point Identification of Visible Light Channel. Before training, collect enough training samples to form a fireworks image set. In order to further simulate the data situation in many real situations and increase the number of training samples, use the data enhancement method to translate, rotate and mirror the flame image to obtain more smoke and fire samples. On the other hand, considering that CNN is used as a classifier, As non-pyrotechnic object, it is also necessary to provide a considerable number of samples, and finally normalize the images. The final data set includes 22565 pictures, of which 17039 are from network public resources and 5526 are from historical videos. Put it into the training set, verification set and test set at the ratio of 2:1:1.

The target detection neural network for fire point recognition uses yolov4, the backbone feature extraction network is cspmarknet53, and the activation function uses the mish activation function:

$$Mish = x \times tanh(ln(1 + ex)) \tag{9}$$

In the feature pyramid part, yolov4 uses the SPP structure and the panet structure. The SPP structure is mixed in the convolution of the last feature layer of cspdarknet53, and the last feature layer of cspdarknet53 is subjected to three times of darknetconv2d_

BN_After leaky convolution, four different scales of maximum pooling are respectively used for processing. The maximum pooled core sizes are 13 × 13, 9 × 9, 5 × 5 and 1 × 1 (1 × 1 means no processing). It can greatly increase the receptive field and isolate the most significant contextual features. Panet can realize repeated feature extraction.

In the feature utilization part, yolov4 extracts multiple feature layers for target detection, and extracts three feature layers, which are located in the middle layer, the middle and lower layers, and the bottom layer.

In the decoding part, yolov4 is to add each grid point to its corresponding x_Offset and Y_Offset, the result after adding is the center of the prediction frame, and then the length and width of the prediction frame are calculated by combining the a priori frame with the height and width. After the final prediction structure is obtained, score sorting and non maximum suppression screening are also performed.

The overall network structure of yolov4 is as follows (Fig. 5):

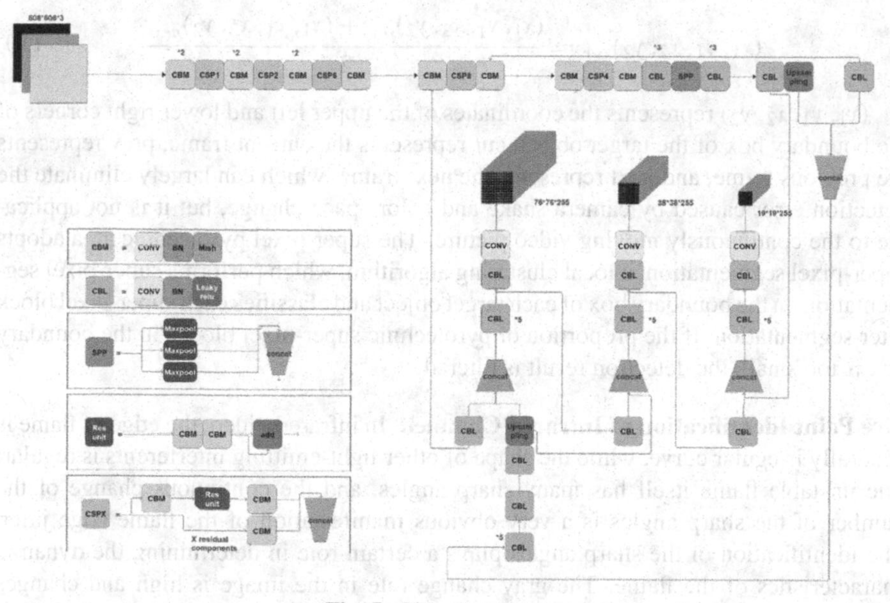

Fig. 5. Yolov4 structure

Input a tensor with the size of n * width * height * 3 to the neural network, indicating that the training set contains n samples, each of which is a three-channel color image of width * height. Take the array as the data type of the input data and labels, define the data and labels, assign values to the data by reading the image file, and then carry out neural network training. The main parameters of the training include input data, label data Batch size and number of training rounds, some common data enhancement methods are added during training.

The trained model is used to identify the fire point of the video stream data, judge whether there is fireworks and control the detection results. The boundary frame of the target can be obtained by using yolov4 model for pyrotechnic target detection. In order to further adjust the detection results on the basis of the model, color space smoke region extraction, context smoothing and super-pixel pyrotechnic region segmentation methods are adopted to post process the boundary frame of the target. Smoke area extraction in color space: extract the smoke area in the boundary box and calculate its proportion according to the characteristics of gray smoke, and filter according to the proportion of the smoke area; There is a great correlation between the current frame of the video stream and the previous and subsequent frames. If the video detects fireworks in the picture of the current frame, but there is no fireworks in the previous and subsequent frames, the detection result of the current frame will be very likely to be false detection. On the contrary, if there is a fireworks detection result in the previous and subsequent frames but there is no previous frame, the current frame will be very likely to have missed detection. In case of false detection and missed detection, The context smoothing method obtains a new detection window by combining the information of the previous and subsequent frames.

$$(x_1, y_1, x_2, y_2)_{cur} = \frac{(x_1, y_1, x_2, y_2)_{pre} + (x_1, y_1, x_2, y_2)_{next}}{2} \tag{10}$$

(x_1, y_1, x_2, y_2) represents the coordinates of the upper left and lower right corners of the boundary box of the target object, cur represents the current frame, prev represents the previous frame, and next represents the next frame, which can largely eliminate the detection error caused by camera shake and color space change, but it is not applicable to the continuously moving video picture; The super-pixel pyrotechnic area adopts super-pixel segmentation, a local clustering algorithm, which performs super-pixel segmentation on the boundary box of each target object and classifies each super-pixel block after segmentation. If the proportion of pyrotechnic super-pixel blocks in the boundary box is too small, the detection result is filtered.

Fire Point Identification of Infrared Channel. In infrared video, the edge of flame is generally irregular curve, while the shape of other light-emitting interferents is regular; The unstable flame itself has many sharp angles, and the continuous change of the number of the sharp angles is a very obvious manifestation of the flame edge jitter. The identification of the sharp angles plays a certain role in determining the dynamic characteristics of the flame. The gray change rate in the image is high and changes rapidly; Compared with the bright objects with interference in flame recognition, the flutter of the outer flame of the flame has randomness and irregularity. Depending on this feature, most of the bright interference sources can be distinguished. The edge change of early fire flame has obvious characteristics, which is different from the edge change of stable flame, and has the characteristics of edge jitter. The infrared channel fire point identification algorithm uses brightness as the main core, and a small number of morphological features including circularity and area size as auxiliary to outline and identify the flame in the video. The specific process is as follows (Fig. 6):

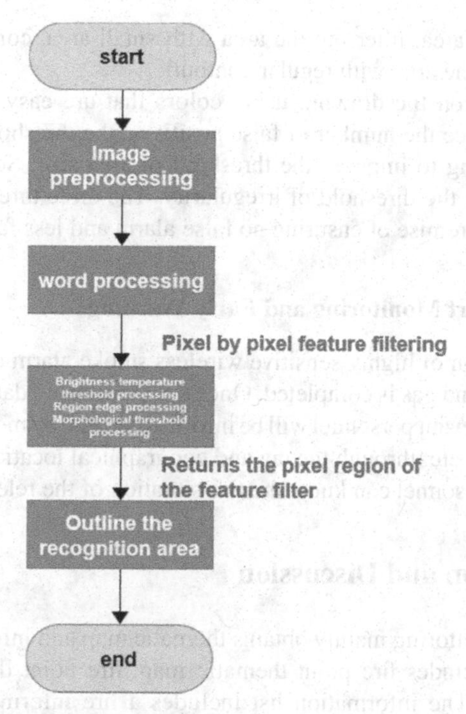

Fig. 6. Infrared video data fire point identification algorithm flow

It mainly includes the processing of the video color channel and the processing of the text information on the video image. Because the format types of streaming media output by different hardware manufacturers are not uniform, it is necessary to conduct unified processing on the video before the subsequent feature analysis. The way is to convert the video channel into a single channel mode with gray information, so as to facilitate the subsequent brightness region division. On the other hand, the processing of the text information is to mask the text on the image. The text information on the image includes date, time and camera position. Information displayed on the gray-scale image in white highlight color, which will have a great impact on the identification of the highlighted flame. It is an important interference body. In the processing of character information, the algorithm refers to the open-source character recognition model, which is based on yolo-v3. It can recognize the character region in complex images, return the quadrangular coordinates of the region matrix, and then mask according to the quadrangular coordinates.

Using the traditional threshold method, the preprocessed infrared image is binarized in black and white, and the obtained area is the preliminarily identified suspected flame area. The edge of the divided area is monitored to obtain the contour of the suspected area, and then the contour area is expanded. The contour area can be used to calculate the morphological characteristics. The main morphological features involved in the algorithm include circularity and area. Calculate the perimeter and area of each area by using the processed contour area, filter out the area with small area, compare the perimeter and area, and remove the area with regular contour.

Finally delineated on the drawing using colors that are easy to observe. Brightness threshold: to reduce the number of false positives, the threshold of binarization is increased; Area filtering to improve the threshold of area size; Screening of regional irregularity to improve the threshold of irregularity. The three thresholds are balanced and debugged on the premise of ensuring no false alarm and less false alarm.

4.3 Inspectors Report Monitoring and Early Warning

Through the installation of highly sensitive wireless smoke alarm device, the real-time monitoring of smoke and gas is completed. Once the monitoring data is found to exceed the alarm value, the relevant personnel will be informed in time by means of mobile phone app push, SMS, email, etc. through the tag and geographical location of the equipment, so that the relevant personnel can know the fire situation of the relevant position.

5 Implementation and Discussion

Finally, forest fire monitoring mainly obtains thematic map and information list results. The thematic map includes fire point thematic map, fire point distribution map and infrared image map; The information list includes a fire information list and a fire information list.

Thematic Map of Forest Fire Monitoring
Superimpose the extracted fire point pixel information on the base map obtained through the above processing, and add auxiliary information such as administrative boundary, occurrence time and place to generate a forest fire monitoring thematic map.

Forest Fire Monitoring Information List
Forest fire monitoring information list includes fire point information list and fire site information list.

The fire point information list includes the name and occurrence time of each province, prefecture and county where the fire point center is located. With reference to the land use classification data, forest land resource data, administrative boundary data and other auxiliary data, the monitoring details and observation details are generated for each fire point information. Monitoring details include: monitoring source, monitoring time, longitude and latitude of the center, name of the province, prefecture and county where the center is located, open fire area, land type, number of pixels, observation times, fire site and verification results. The observation details are the historical records of the fire point, including the monitoring time, monitoring source and pixel number of the fire point.

The fire information list includes the name of the province, prefecture and county where the fire center is located, the occurrence time and the fire treatment status. Similar to the fire information list, the monitoring details and monitoring records are generated for each fire information by referring to the land use classification data, forest land resource data, administrative boundary data and other auxiliary data. The monitoring details include the discovery (fire site number, monitoring source, update time, longitude and latitude, detailed address, open fire area, land type, original discovery time, original ownership, etc.), early warning status, alarm status, verification results and settlement status. Monitoring records include fire point observation frequency, monitoring time, monitoring source and number of pixels.

System Related Screenshots

See Figs. 7, 8, 9 and 10.

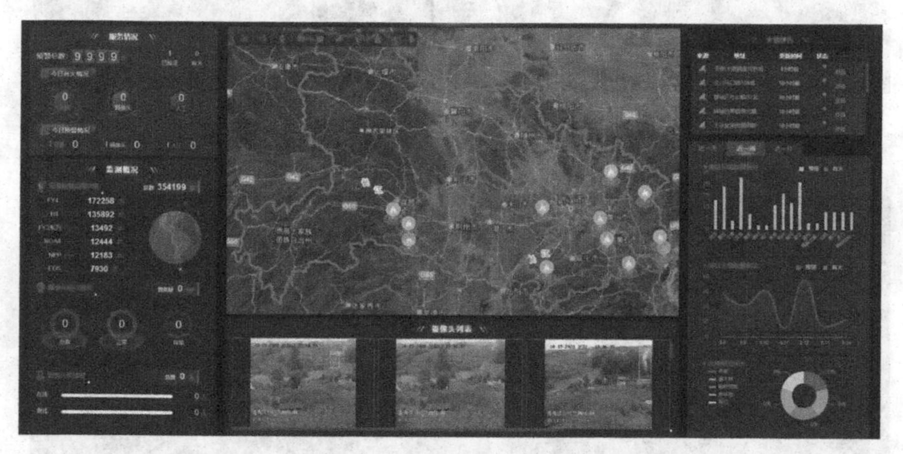

Fig. 7. System homepage

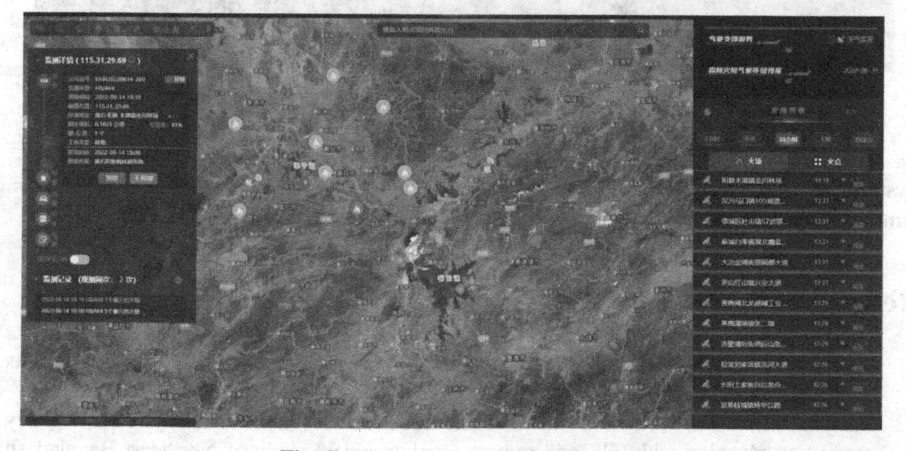

Fig. 8. Fire point monitoring details

Fig. 9. Video monitoring details

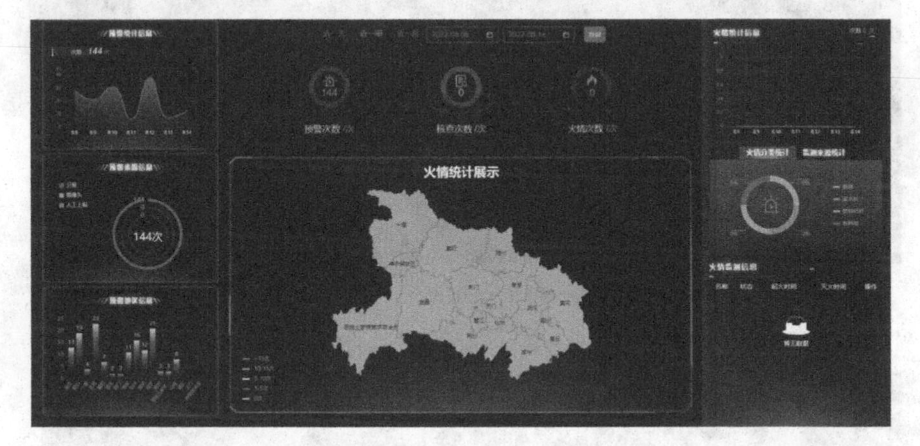

Fig. 10. Data statistics

System Screen Recording Details. https://v.youku.com/v_show/id_XNTg5MzY1NTE3Mg==.html?x&sharefrom=android&sharekey=fd148c595c18e52af6008f24121407333.

References

1. Barrile, V., Bilotta, G., Fotia, A., et al.: Integrated GIS system for post-fire hazard assessments with remote sensing. Copernicus GmbH (2020)
2. Rahman, M., Chen, N., Islam, M.M., et al.: Location-allocation modeling for emergency evacuation planning with GIS and remote sensing: a case study of Northeast Bangladesh (2021)
3. Bourjila, A., Dimane, F., Nouayti, N., et al.: Use of GIS, remote sensing and AHP techniques to delineate groundwater potential zones in the Nekor Basin, Central Rif of Morocco. In: GEOIT4W-2020: 4th Edition of International Conference on Geo-IT and Water Resources 2020, Geo-IT and Water Resources 2020 (2020)

4. Chen, H., Duan, S., Ge, X., et al.: Multi-temporal remote sensing fire detection based on GBDT in Yunnan area. In: 2020 2nd International Conference on Machine Learning, Big Data and Business Intelligence (MLBDBI) (2020)
5. Kumar, R.: assessing fire risk in forest ranges of Guntur district, Andhra Pradesh: using integrated remote sensing and GIS. Int. J. Sci. Res. (IJSR) **3**(6), 1328 (2021)
6. Boselli, A., Sannino, A., D'Emilio, M., et al.: Aerosol characterization during the summer 2017 huge fire event on Mount Vesuvius (Italy) by remote sensing and in situ observations. Remote Sens. **13**(10), 2001 (2021)
7. Kanakaraja, P., Sundar, P.S., Vaishnavi, N., et al.: IoT enabled advanced forest fire detecting and monitoring on Ubidots platform. Mater. Today Proc. **46**, 3907–3914 (2021)
8. Singh, H., Shukla, A., Kumar, S.: IoT based forest fire detection system in cloud paradigm. IOP Conf. Ser. Mater. Sci. Eng. **1022**(1), 012068 (2021)
9. Singh, R., Gehlot, A., Shaik, V.A., et al.: Forest 4.0: digitalization of forest using the Internet of Things (IoT). J. King Saud Univ. Comput. Inf. Sci. (2021)
10. Sun, L., Wan, L., Wang, X.: Learning-based resource allocation strategy for industrial IoT in UAV-enabled MEC systems. IEEE Trans. Industr. Inform. **17**(7), 5031–5040 (2020)
11. Kumar, D., Kumar, A.K., Majeeth, A., et al.: Forest fire recognition and surveillance using IoT. J. Adv. Res. Dyn. Control Syst. **12**(5), 1085–1089 (2020)
12. Gaitan, N.C., Hojbota, P.: Forest fire detection system using LoRa technology. Int. J. Adv. Comput. Sci. Appl. **11**(5) (2020)

Research on the Pricing Method of Power Data Assets

Ting Wang[1], Lina Jiang[1], and Xiaohua Wang[2](✉)

[1] Big Data Center of State Grid Corporation of China, Beijing 100033, China
[2] Beijing Sgitg Accenture Information Technology Center Co., Ltd., Beijing 100052, China
wangxiaohua01@sgitg.sgcc.com.cn

Abstract. Today, when the value of data is prominent, higher requirements are placed on the transaction of power data assets, both inside and outside the enterprise. At present, the value of data assets within the company is unclear, and there is no unified data asset pricing method to standardize pricing. The pricing of power data assets is an important basis for power data asset transactions. In order to solve the above problems, based on the analysis of power data assets, combined with common data asset valuation methods, this paper proposes production pricing method. It is hoped that through the pricing of power data assets, the transaction process of power data assets will be promoted and the development of the digital economy will be promoted.

Keywords: Power · Data asset · Pricing · Principle · Method

1 The Introduction

State Grid's informatization started early. After the informatization construction from the "Eleventh Five-Year Plan" to the "Thirteenth Five-Year Plan" period, it has built ten application systems with two-level deployment, which has laid a good foundation for the company's digital development. The company has the world's largest integrated enterprise group information system, accumulated massive data resources, initially realized the integration and unified management of group data, built an integrated business application system and typical scenarios, and effectively supported the company's main business data need.

At present, the economic and social value contained in power big data has been gradually discovered and realized value-added, which has become an important asset of enterprises. Power data asset trading has gradually become the focus of enterprises. The design and construction of a scientific power data asset pricing system is of great significance to clarify the attributes of power data assets, formulate a unified pricing mechanism, and promote the healthy development of the power data market.

For enterprises, promoting data asset pricing can effectively improve the operational efficiency and management capabilities of each business line of the enterprise. Regular pricing work can effectively assist management in analyzing the high correlation between data asset value and enterprise value, so as to explore high value density data,

B. Hu et al. (Eds.): BigData 2022, LNCS 13730, pp. 52–60, 2022.
https://doi.org/10.1007/978-3-031-23501-6_6

formulate or revise business development goals and strategies. In addition, data assets with accurate quantitative value can effectively solve the communication barriers in performance understanding and digital operation of various departments of the enterprise, reduce the communication cost of the enterprise, improve the operation efficiency, and promote the overall development of the enterprise in a healthy direction [1].

In order to solve the problem of power data asset pricing, based on the understanding and analysis of domestic and foreign research, and on the basis of optimizing the classical data asset pricing theory, based on the status quo of the State Grid Corporation of China, this paper proposes a power data asset I hope to take this opportunity to conduct in-depth exchanges and discussions with academia, scientific research units and industry insiders, and lay the foundation for building a complete power data asset valuation framework system and opening up a complete cycle chain of the power data asset market.

2 Research Status at Home and Abroad

In terms of data pricing, the earliest foreign report on the value of data assets was "Why and how to measure the value of your information assets" (Chinese name: How to measure the net value of your information assets) proposed by Gartner in 2015. Recently, a total of 8 data asset valuation methods for reference have been proposed, including IVI, BVI, PVI, CVI, MVI, EVI and other data valuation methods such as monetization and non-monetization. At the same time, DAMA-DMBOK (Chinese name: Data Management Knowledge Theory System Guide) also proposes three data asset evaluation methods from the perspectives of the characteristics, influencing factors, and business models of data asset valuation objects [2].

2.1 Current Status of Foreign Research

With the deepening of people's understanding of the concept of big data, the idea of "Data as a Service (DaaS)" continues to prevail. In the DaaS environment, researchers have discussed the pricing model of data products, which are mainly divided into the following three types: company subscription, capacity-based pricing, and data-type-based pricing [3].

(1) Company subscription: Data service providers charge corresponding fees for providing data product subscriptions to companies or individuals.
(2) Pricing based on capacity: Data service providers charge users based on the amount of data user access through application programming interfaces (APIs).
(3) Pricing based on data type: Pricing is based on data type or attribute division.

The data pricing model in the context of DaaS provides a new idea for big data pricing. On the basis of the above pricing model, researchers have further studied and proposed some emerging data pricing methods.

2.2 Domestic Research Status

With the advancement of data processing technology, more and more data are mined and analyzed for corporate decision-making, which improves operational efficiency. But for big data, the data itself has some characteristics, such as no physical form, value uncertainty, can be used as non-monetary long-term assets, transferable use rights, etc. Therefore, some researchers advocate the use of big data as an intangible asset for pricing, and apply the asset pricing method to data pricing. Application in pricing.

In addition to using traditional asset pricing methods to price data assets, some emerging pricing methods that combine the characteristics of big data have emerged. Zhang Zhigang et al. analyzed the composition of the value of data assets from the cost and application of data assets. The data cost consists of construction costs and operation and maintenance costs. The construction costs include labor costs, material costs, and indirect costs. The operation and maintenance costs include business operation costs and technical operation and maintenance costs., Use effect evaluation of four indicators; based on the research on data value composition, Wang Jianbo proposed that artificial intelligence methods such as neural networks can be used to discuss the pricing of data assets, and by inputting data value indicators, output data intrinsic value or market value Equivalent value indicators to realize the pricing of data assets [4]; Qu Lili and others pay attention to the value assessment of alliance data assets, and use the real option-based data asset assessment method to calculate the value of alliance targets and execution data assets through the importance of each alliance member enterprise, and use improved The B-S model evaluates consortium data assets [5].

3 Data Asset Valuation Methods

Since data assets do not have a physical form, they are usually evaluated by analogy to intangible assets. In order to guide asset appraisal institutions and their asset appraisal professionals to carry out data asset appraisal business, the China Asset Appraisal Association has formulated the "Guidelines for Asset Appraisal Experts No. 9 - Data Asset Appraisal", which points out that the appraisal methods of data asset value include cost method, there are three basic methods of income method and market method and their derivatives [6–8].

3.1 Cost Method

The cost method is essentially the collection of data costs, mainly including the costs incurred in the process of data generation and application, such as labor, material costs and other construction costs during the construction period, collection, processing costs and other operation and maintenance costs. Operation and maintenance costs. The cost method applies to data that has no apparent market value or is generating a market.

Advantages: Collect all the expenses generated by the data, which is easy to grasp and operate, and facilitate financial processing. Since the accounting measurement of production cost adopts cost value accounting, it is suitable for data collection and value summary of the balance sheet.

Limitations: It cannot reflect the benefits that data assets can generate, and the corresponding costs of data assets are not easy to distinguish. Some data assets do not have corresponding direct costs, and it is difficult to estimate the allocation of indirect costs.

3.2 Income Method

Income method is a valuation method to obtain the present value of economic benefits generated in the future based on the expected application scenarios of target assets. The premise of application is that the future expected return of the asset under evaluation can be predicted and can be measured in currency.

Advantages: It can reflect the value of data assets more realistically and accurately. When the expected income of data assets can be estimated, the corresponding data value is also relatively easy to predict.

Limitations: The amount of benefits and risks are easily affected by subjective judgment factors and are difficult to predict accurately.

3.3 Market Law

The market method evaluates the value of data assets by analogy with data transaction prices. The premise of application is that there is an open and active trading market for the underlying asset or similar assets, and the transaction price is easy to obtain.

Advantages: Due to the fairness of the reference market price, it is easier to be accepted by both buyers and sellers.

Limitations: The requirements for market transaction conditions and transaction volume are relatively strict, the data assets have certain confidentiality, and the transaction types are relatively simple, so it is difficult for evaluators to collect the market price of data transactions.

4 Power Data Asset Pricing

4.1 Characteristics of Power Data Assets

Power data assets have their special business attributes, management attributes and technical attributes, which are embodied in "5V" characteristics, including Value, Volume, Frequency Velocity, Real Veracity, and Variety.

(1) In terms of value, power data assets have the characteristics of high value density and strong professional attributes. Power data directly reflects the operating status of the power grid, the power consumption behavior of industrial customers, and the operation of enterprises. It is a twin mirror image of the social production and living status from the perspective of energy, and has unique value advantages.
(2) In terms of capacity, power data assets have the characteristics of large collection scale and fast growth rate. The number of power data collection objects is huge, involving billions of grid equipment and billions of upstream and downstream entities. The service objects cover all industries and fields, and the total amount of data has reached PB level.

(3) In the frequency dimension, power data assets have the characteristics of high real-time and strong continuity. Electricity production, transmission, distribution, and consumption are completed at the same time. Electricity data represents the whole process of energy flow, and the collection frequency of some data reaches the millisecond or minute level, which is highly real-time.
(4) In the real dimension, power data assets have the characteristics of high accuracy and strong sensitivity. Power supply has high requirements for safety and stability. Through the holographic perception of power grid production and operation status through terminal equipment, the real and comprehensive acquisition of data is realized, which effectively guarantees the accuracy and reliability of data. The company's service subjects are diverse, involving national economy and people's livelihood, energy security and customer privacy, and are highly sensitive.
(5) In various dimensions, power data assets have the characteristics of large business span and many data types. Electric power data covers various fields of enterprise management such as human, property, and property, covering all majors in power grid production such as construction, production, and marketing, and involves various business sectors such as finance, industry, and international business. The data presents significant multi-source and heterogeneous characteristics.

4.2 Division of Power Data Assets

According to functional applications, power data assets can be divided into real-time exchange data, offline data packets, models, algorithms, and computing power. In fact, the delivery of data products such as real-time exchange data and offline data packages is mostly standardized data products, while models and algorithms are more customized data products.

4.3 Pricing Principles

The characteristics of power data assets also determine their pricing diversity, combination and multiple times [9].

(1) Diversity
Power data asset trading is different from general industrial product trading. There are various forms, such as data, API, analytical products, solutions, etc., and its application scenarios, users, trading regulations and other factors are constantly changing, which requires specific the types of power data assets and specific scenarios, objects and other factors are specifically priced.
(2) Repeatedly
Power data assets can be traded and used multiple times. The specific needs of each transaction are different, and the accuracy requirements for the cost of construction are also different, which determines that the pricing method has multiple characteristics. This requires that the power data assets need to be re-priced before each transaction in combination with different application scenarios, objects of use, real-time status and other factors of the data, so as to accurately estimate the value of the current power data assets.

(3) Combination
The pricing of power data assets includes all expenses from data planning, collection and verification, storage, data management system construction and operation and maintenance, data analysis functions and API custom development. The production process is decomposed into the basic components that can calculate the cost, and then gradually aggregated, in order to accurately calculate the entire cost price of the entire data product.

(4) Volatility
The value of power data assets is affected by many different factors, including technical factors, data capacity, data value density, business models of data applications, and other factors, which change over time, resulting in the volatility of data asset value.

4.4 Pricing Method

On the basis of the cost price method, the factors affecting the realization of the value of power data assets and market supply and demand factors are comprehensively considered to price power data assets.

The power data asset pricing method is based on the price (tax included) calculated based on the modified cost price method, and its expression is as follows:

$$V_d = C \times (1 + R) \times L \tag{1}$$

in:

C—total cost;
R—cumulative profit margin;
L—data usage trend coefficient.

(1) Data asset cost C

$$C = \left[C_0 \cdot \left(\sum_{i=1}^{n} q_i \right)^{\alpha} + \sum_{i=1}^{7} C_i \right] \times S \tag{2}$$

Among them, C_0 is the acquisition cost of data assets, q_i is the quality factor, $C_1 \ldots C_7$ are the costs incurred in data asset processing, storage, security, maintenance/update, product development, management, sales, etc., S is the safety factor, α In order to consider whether to consider the value coefficient of data quality, when the quality coefficient needs to be considered, $\alpha = 1$, data quality, that is, the degree to which the characteristics of the data meet the requirements when used under specified conditions; When purchasing the data that has been sorted and analyzed, the above-mentioned quality factors have been considered in the transaction price, so there is no need to additionally consider the adjustment of the quality coefficient ($q_1 \ldots q_n$) in the calculation, at this time $\alpha = 0$.

The specific components of the acquisition cost C_0 of data assets include (Table 1):

Table 1. Sources of acquisition cost of data assets

Raw data assets	External acquisition	Purchase price and taxes
	Internal collection	Collection staff cost
		Purchasing staff cost
		The cost of purchasing terminal equipment
		Procurement system cost
		Bad Data Obsolescence Costs
		Other acquisition costs

The processing cost C_1 of data assets includes:

- Processing personnel costs
- Processing system cost
- Other processing costs

Storage cost of data assets C_2:

- Storage equipment costs
- Other storage costs

Security cost of data assets C_3, maintenance and update cost C_4:

- Data asset maintenance personnel costs
- Data asset maintenance system cost
- Updating staff costs
- Updating system cost
- Security maintenance costs

Development cost of data asset products C_5:

- Developer costs
- Development system cost
- other costs

Management costs of data assets C_6:

- Management staff costs
- Manage system costs
- other costs

Cost of sales of data assets C_7:

- Salesperson costs
- Cost of selling the system
- other costs

(2) qi quality coefficient

If the data purchased by the enterprise from outside has not been cleaned and verified, the quality factor ($q_1 \ldots q_n$) needs to be considered when calculating the data acquisition cost (C_0). The quality coefficient ($q_1 \ldots q_n$) is the sum of the evaluation results of each inspection dimension, generally ≤ 1 (Table 2).

Table 2. Data quality coefficient calculation table

Data asset pricing object	Inspection dimension	Inspection standard	Inspection result	Data quality coefficient α (≤1)
	Accuracy			
	Integrity validity			
	Uniqueness			
	Accessibility			

(3) Cumulative profit rate R

$$R = \left[\sum\nolimits_{T=1}^{n} R_T - \sum\nolimits_{T=1}^{n} C_T\right] \div \sum\nolimits_{T=1}^{n} C_T \tag{3}$$

The cumulative profit margin R of the data asset establishes the profit margin so far for the data asset. R_T is the profit in a certain year and C_T is the cost in a certain year. n is the age to date for the data asset established.

(4) Data usage trend coefficient L

$$L = (L_T - L_{T-1})/L_{T-1} \tag{4}$$

$$L_T = \frac{\text{Access to data assets in the current year}}{\text{The total amount of data assets in the current year}} \tag{5}$$

The data usage rate trend coefficient L is the ratio of the difference between the data usage rate of the current year and the data usage rate of the previous year and the data usage rate of the previous year, which reflects the development trend of the proportion of data used in the data assets to the total assets. It reflects the trend of the proportion of data involved in the operation and transaction of data assets.

5 Summary and Outlook

With the prominence of data value, both inside and outside the State Grid Corporation, higher requirements have been placed on the transaction of power data assets, and the current status quo has been unable to meet the growing demand for data transactions both inside and outside the enterprise. Therefore, it is necessary to promote reasonable pricing of power data assets on the basis of clarifying the input, use, and transaction requirements of internal and external data of the enterprise, which is an important part of realizing the value of power data assets. It is expected that with the resolution of various transaction problems, the scale of power big data transactions can rapidly expand, effectively promoting the development of the digital economy.

References

1. PricewaterhouseCoopers. White Paper on Prospective Research on Data Asset, pp. 13–15 (2021)
2. Chen, Y.: Research on big data pricing method based on utility. Chongqing Jiaotong University, pp. 2–8 (2020)
3. Luo, Z.: Research on data pricing method based on VCG and Rubinstein model. Harbin Institute of Technology, pp. 1–14 (2020)
4. Zhang, Z., Yang, D., Wu, H.: Research and application of data asset valuation model. Mod. Electron. Technol. 44–47 (2015)
5. Qu, L., Ma, Z., Zhang, S.: A survey on pricing of big data products. Science-Technology and Management, pp. 105–110 (2018)
6. China Assets Appraisal Association. Asset Appraisal Expert Guidelines No. 9 - Data Asset Appraisal (2019). http://law.esnai.com/print/195061
7. Lookout, a think-tank. Commercial bank data asset valuation white paper, pp. 22–57 (2021)
8. Xiong, Q., Tang, K.: Research progress on the rights, trading and pricing of data elements. Econ. Perspect. **2**, 151–153 (2021)
9. China Southern Power Grid Co., Ltd. Data asset pricing method and fee standard (trial implementation), pp. 3–22 (2021)

Research and Design on the Confirmation Method of Power Data Assets

Jiang Lina[1,2], Wang Ting[1], and Wang Xiaohua[2](✉)

[1] Big Data Center of State Grid Corporation of China, Beijing 100033, China
[2] Beijing Sgitg Accenture Information Technology Center CO., LTD., Beijing 100052, China
wangxiaohua01@sgitg.sgcc.com.cn

Abstract. Today, when data sharing is becoming more and more important, both internal and external enterprises have put forward higher requirements for the open sharing of power data assets. It is difficult to ensure that big data analysis and data services are carried out under the premise of unclear ownership of power data assets. The security of power data assets is not conducive to the safe development of the power industry. In order to solve the above problems, this paper classifies the power data assets based on the analysis of the power data assets, and divides the property rights of the power data assets according to the power data flow process, and proposes the confirmation of the power data assets. Method. It is hoped that through the confirmation of power data assets, the realization of open sharing of power data assets will be promoted.

Keywords: Data asset · Power · Data property right · Data right · Data asset confirmation · Data classification

1 The Introduction

State Grid's informatization started early. After the informatization construction from the "Eleventh Five-Year Plan" to the "Thirteenth Five-Year Plan" period, it has built ten application systems with two-level deployment, which has laid a good foundation for the company's digital development. The company has the world's largest integrated enterprise group information system, accumulated massive data resources, initially realized the integration and unified management of group data, built an integrated business application system and typical scenarios, and effectively supported the company's main business data need.

At present, the economic and social value contained in power big data has been gradually discovered and realized value-added. It has become an important asset of the enterprise and plays an important role in promoting the development of the enterprise. In order to fully utilize the power data assets, State Grid Corporation of China continues to promote data sharing. However, there are still many problems in the open sharing of power data, which is not conducive to realizing the digital transformation needs of power enterprises. For example, the foundation of data sharing is not yet solid, the sharing mechanism is not perfect, the data use authorization process needs to be further

B. Hu et al. (Eds.): BigData 2022, LNCS 13730, pp. 61–69, 2022.
https://doi.org/10.1007/978-3-031-23501-6_7

optimized, data privacy protection and compliance management need to be strengthened, and data risk prevention and control capabilities need to be improved urgently. The opening of power data assets still needs to be explored and promoted, the data opening mode, standards and process need to be further studied, and the data opening boundary and strategy need to be clarified. Secondly, the current open sharing of power data assets is more inclined to be within the enterprise, and it is impossible to exchange a large amount of data with external related institutions, enterprises, and the public.

Today, when data sharing is becoming more and more important, both internal and external enterprises have put forward higher requirements for the open sharing of power data assets. It is difficult to ensure that big data analysis and data services are carried out under the premise of unclear ownership of power data assets. The security of power data assets will cause potential problems such as leakage of key power information, which is not conducive to the safe development of the power industry. To realize the fully open sharing of power data assets, it is necessary to clarify the data ownership of power data assets.

In order to solve the above problems, based on the analysis of power data assets, this paper divides the property rights of power data assets, and proposes a method for confirming the rights of power data assets. In the process of confirming the rights of power data assets, we try to solve two problems: (1) clarify the property rights content of power data assets, that is, what rights can be allocated; (2) the grant of property rights of power data assets, that is, how to divide who owns what rights according to the situation.

2 Analysis of Power Data Assets

2.1 Power Data Asset Concept

Power data assets refer to those that are legally owned, controlled or used by power enterprises through procurement, collection, and production, and are formed by matters such as power supply and electricity consumption, and are expected to bring economic benefits to power enterprises. Data resources recorded electronically.

2.2 Power Data Asset Characteristics

Power data assets have their special business attributes, management attributes and technical attributes, which are embodied in the "5 V" characteristics, including value, capacity, frequency, authenticity and variety.

(1) In terms of value, power data assets have the characteristics of high value density and strong professional attributes. Power data assets directly reflect the operation status of the power grid, the power consumption behavior of industry customers, and the operation of enterprises. They have strong professional attributes of power, and are the "barometer" and "wind vane" of economic and social operation, with unique value advantages.

(2) In terms of capacity, power data assets have the characteristics of large collection scale and fast growth rate. The number of power data collection objects is huge, involving billions of grid equipment and billions of upstream and downstream entities. The service objects cover all industries and fields, and the total amount of data has reached PB level.
(3) In the frequency dimension, power data assets have the characteristics of high real-time and strong continuity. Electricity production, transmission, distribution, and consumption are completed at the same time. Electricity data represents the entire process of energy flow. The collection frequency of some data reaches the millisecond or minute level, which has strong real-time and continuity.
(4) In the real dimension, power data assets have the characteristics of high accuracy and strong sensitivity. Power supply has high requirements for safety and stability. Through the holographic perception of the power grid production and operation status through terminal equipment, the data is true, accurate and reliable. At the same time, as the business involves the national economy and people's livelihood, energy security and customer privacy, power data assets are highly sensitive.
(5) In various dimensions, power data assets have the characteristics of large business span and many data types. Power data assets cover all fields of enterprise management, covering construction, production, marketing and other power grid production majors, involving finance, industry, international and other business sectors, involving various structured, unstructured, semi-structured data and geospatial Vector data, showing significant multi-source heterogeneity.

2.3 Classification of Power Data Assets

State Grid has formed a complete data classification management framework based on the industry's data classification and its years of practice. The purpose of classifying data by State Grid is to adopt different management strategies for data with different characteristics, in order to maximize the value of data assets.

State Grid divides data into: enterprise internal data and external data, structured data and unstructured data, metadata, customer data, enterprise data, and social data. Among them, structured data is further divided into reference data, observation data, master data, transaction data, report data and rule data.

2.4 Power Data Asset Classification

According to the affected objects and the degree of impact after the security of power data assets is damaged, the security level of data assets is divided into 5, 4, 3, 2, and 1 from high to low, which generally have the following characteristics [1–3]:

– The characteristics of level 5 data assets are as follows:

• Important data assets, usually mainly used for key business use of important core nodes in the data transaction process of power enterprises, are generally disclosed to specific personnel, and are only accessed or used by those who must know.

- After the security of data assets is destroyed, it will have an impact on national security or have a serious impact on public rights and interests.

– The characteristics of level 4 data assets are as follows:

- Data assets are usually mainly used for important business use of important core node institutions in the data transaction process of power enterprises. They are generally disclosed to specific persons, and are only accessed or used by those who must know.
- After the security of data assets is destroyed, it will have a general impact on public rights and interests, or have a serious impact on personal privacy or the legitimate rights and interests of enterprises, but will not affect national security.

– The characteristics of level 3 data assets are as follows:

- Data assets are used for critical or important business use by power companies, generally disclosed to specific persons, and only accessed or used by those who must know.
- After the security of data assets is destroyed, it has a slight impact on public rights, or has a general impact on personal privacy or the legitimate rights and interests of enterprises, but does not affect national security.

– The characteristics of Level 2 data assets are as follows:

- Data assets are used for general business use in the data transaction process of power enterprises, and are generally disclosed to restricted objects, which are usually internally managed and should not be widely disclosed.
- After the security of data assets is destroyed, it will have a slight impact on personal privacy or the legitimate rights and interests of enterprises, but will not affect national security and public rights and interests.

– The characteristics of Level 1 data assets are as follows:

- Data assets are generally available to the public or to be known and used by the public.
- Information that is voluntarily disclosed by the subject of personal information.
- After the security of data assets is destroyed, it may not affect personal privacy or the legitimate rights and interests of enterprises, or only slightly affect national security and public rights and interests.

2.5 Data Flow Process

At present, the State Grid data asset system is built around the data center. The data center gathers data from various professional business systems, business centers, IoT management platforms, and external data sources, cleans, transforms, and integrates data according to a unified model, realizes data standardization, integration, and labeling unified storage, and provides data resources, data analysis and data labeling and other sharing services to improve the ability of data convenient application.

3 Division of Property Rights of Power Data Assets

The property rights of power data assets are composed of various rights. The property rights of power data assets can be divided and continuously expanded with the evolution of social and economic life. The purpose of setting the property rights of power data assets is to reasonably protect the privacy of individuals and consumers, and at the same time encourage power companies to collect data and fully develop and utilize data elements.

The general idea of the property rights setting of power data assets is to emphasize that the collective rights and interests are not infringed and the shared benefits are maximized for enterprise data and social data, and to strengthen the privacy protection of personal information for user data. Based on this, the property rights system for constructing power data assets is divided as follows (Fig. 1).

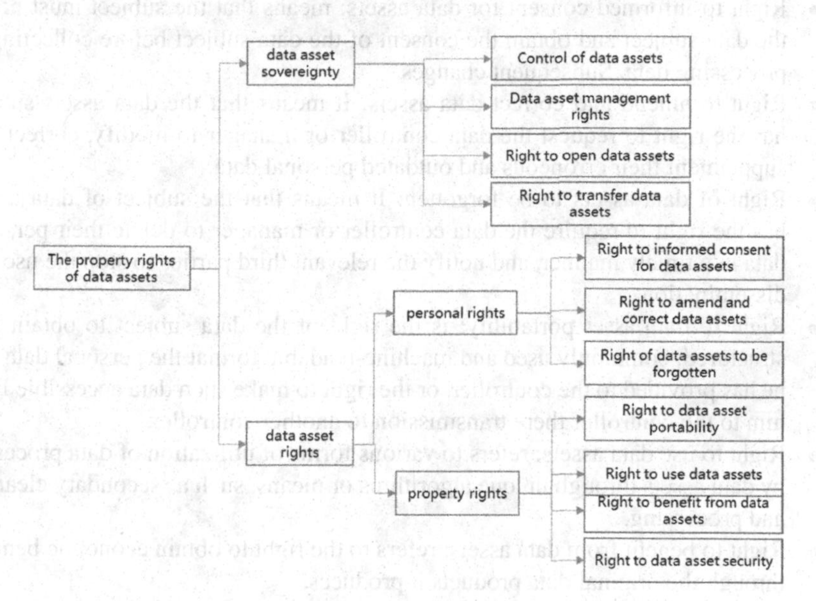

Fig. 1. Division of property rights of power data assets

The property rights of data assets are divided into two categories: data asset sovereignty and data asset rights [4, 5].

(1) The connotation of data asset sovereignty is the exercise of jurisdiction over data assets. The sovereignty of data assets includes the right of control, management, openness and transfer.

- Control of data assets: Take effective measures to protect the security, authenticity and integrity of data assets to prevent data assets from being tampered with, forged, and leaked. Rights to data not expressly bound by law or other contract can be enforced at their own will.
- Data asset management rights: administer the entire life cycle of production, processing, and circulation of data assets.
- Right to open data assets: Open and share the data assets you have mastered as needed.
- Right to transfer data assets: The right to transfer one's legitimate interests or rights in original data to others.

(2) The connotation of data asset rights is the privacy protection and use of data assets. Data asset rights include personal rights and property rights. Personality rights of data assets also include the right to informed consent, the right to amend and rectify, and the right to be forgotten. The property rights of data assets include the right to portability, the right to use, the right to benefit, and the right to security.

- Right to informed consent for data assets: means that the subject must notify the data subject and obtain the consent of the data subject before collecting or processing data. Subsequent changes.
- Right to amend and correct data assets: It means that the data assets subject has the right to request the data controller or manager to modify, correct and supplement their erroneous and outdated personal data.
- Right of data assets to be forgotten: It means that the subject of data assets has the right to require the data controller or manager to delete their personal data in a timely manner, and notify the relevant third parties to stop the use and dissemination.
- Right to data asset portability: is the right of the data subject to obtain in a structured, commonly used and machine-readable format the personal data that he has provided to the controller, or the right to make such data accessible from him to the controller there transmission to another controller.
- Right to use data assets: refers to various forms of utilization of data processed by data assets through unique algorithms or means, such as secondary cleaning and processing.
- Right to benefit from data assets: refers to the right to obtain economic benefits through the original data products it produces.
- Right to data asset security: The right to be free from deprivation is a right to prevent data from being illegally intruded, known, collected, utilized and disclosed by others.

4 Power Data Assets Confirmation Scheme

4.1 Confirmation Principle

Data confirmation makes power data assets controllable, which is conducive to speeding up the circulation of data elements. By determining the property rights of power data, the data holders' expectations for future data use rights and revenue rights are stabilized, the risk of data disputes is reduced, and the enthusiasm of power companies to participate in the data element market is improved.

At present, the exploration of the right to confirm the power data assets is still in its infancy, and it will take a certain amount of time for the actual implementation. In the early stage of power data asset confirmation, we can start from the construction of data principles, based on the nature of data, grasp the direction of rights confirmation, and then formulate rights confirmation methods [4–8].

The first is the principle of division of property rights. The purpose of confirming the ownership of power data assets is to achieve the incentive compatibility of different stakeholders, that is, to balance the rights and interests of each participant in the power data value chain, and to achieve data-driven economic development based on the reasonable protection of user privacy. Therefore, the confirmation of power data assets needs to solve the ownership of rights and interests attached to the data rather than pure ownership. The core is to determine which interests should be protected, that is, the protection of data rights should be interests rather than ownership. The core function of ownership is to clarify the exclusivity of property rights, and property rights are different from ownership and have a broader scope than ownership. Establishing a power data asset confirmation property rights framework is more conducive to realizing data asset and promoting data transaction circulation than clarifying data ownership.

The second is the principle of subject classification. According to the different data subjects of power data assets, the data is divided into three categories: user data, enterprise data, and social data.

- User data refers to data that can identify the identity of a natural person or data generated due to the behavior of a natural person. User data has obvious sensitive privacy characteristics. If the user expressly agrees to the enterprise to collect his personal information, then when the enterprise obtains economic or other benefits through the data, the individual should also partially enjoy the right to benefit from the data. This right to benefit is not necessarily reflected in cash or currency. Users provide some free value-added services in addition to currency, and the sharing methods can be flexible and diverse.
- Enterprise data refers to all kinds of data generated or legally acquired by enterprises in production, operation and management activities. Enterprise data consists of enterprise main data and enterprise data authorized by users. Enterprises have data property rights to the main data, and have partial rights to the data derived from users. After obtaining the user's consent, on the premise of not infringing on the privacy of personal data, the next step of development and utilization can be carried out after desensitization and declassification.
- Social data includes all kinds of data and derived data collected by the government and public agencies in accordance with the law in carrying out activities. Governments and public institutions have data property rights to social data. Social data often involves social public interests, and the granting of its ownership should focus on the attribute of social factors. The government and public institutions perform their duties with the support of public finances, and have the responsibility to provide public goods and

create public value to the society. Therefore, social data should be openly shared to the society without involving personal privacy security and national security.

The third is the principle of optimal distribution of value output. According to modern property rights theory, in modern production activities, the ownership of ownership should fully consider the party with the largest marginal contribution in cooperative output. In the field of power data asset ownership confirmation, the determination of power data ownership should also fully encourage the subjects corresponding to important links with high marginal cost and more prominent value creation in the entire power data industry chain. After considering the modern property rights theory, the ownership of power data assets should be clarified before and after all unexpected products in the data industry, so as to effectively protect the interests of investors and avoid the phenomenon of "robbery" of income, which can encourage more Many investors have joined the various industrial chains of power data, so as to achieve optimal output results.

4.2 Power Data Assets Confirmation Scheme

(1) Sharing within the enterprise

Within an enterprise, data assets are divided into different data themes, which include various data entities and attributes.

Data asset ownership confirmation is generally based on the data asset catalog, identifying data entities in various professional fields, establishing a CRUD matrix between the data entity and all parties in the organization, and determining the property rights of data assets.

Generally speaking:

Customer data: Whoever produces the data has data asset sovereignty, data asset personality rights, and data asset property rights. Among them, the right to security is jointly guaranteed by customers and enterprises.

Enterprise data: Whoever produces the data has the sovereignty and personality rights of data assets. The property rights of data assets belong to enterprises and related business departments.

Social data: Whoever produces the data has data asset sovereignty, data asset personality rights, and data asset property rights. General social data belongs to outsourced data, and the above rights belong to the enterprise.

When sharing within an enterprise, what is shared is the right to use and portability in the property rights of data assets, the right to security is jointly guaranteed by the customer and the enterprise, and the right to benefit belongs to the customer or business department.

(2) External transactions of enterprises

When an enterprise conducts data transactions with external parties, the property rights of data assets change differently depending on the transaction form.

When trading data: After desensitization and other processing are performed according to the requirements of the data security level, with the completion of the transaction process, the property rights of the data change, and the sovereignty of the data assets and the rights of the data assets are transferred. According to the agreement between the two parties, the acquirer has the property rights of the data assets, the rights to the data assets, the right to use the data assets, etc. At the same time, both parties need to ensure the security of data assets. It should be noted that the property rights of the original data assets of the deliverer have not changed.

When trading data APIs and services: only the portability and usage rights of data assets change. At the same time, both parties need to ensure the security of data assets.

5 Summary and Outlook

With the increasing importance of data sharing, both within and outside the State Grid Corporation, higher requirements have been placed on the open sharing of power data assets, and the current status of open sharing applications can no longer meet the growing data demands inside and outside the enterprise, is not conducive to the realization of new power system construction. Therefore, on the basis of clarifying the subject of data rights inside and outside the enterprise, focusing on the right to use data, combined with the hierarchical and classified management of data, a set of data rights confirmation schemes are proposed for both inside and outside the enterprise, in order to ensure the security of power data assets and fully realize the power The internal sharing and external opening of data assets provide support, thereby simplifying the internal approval process and data acquisition process of the enterprise, and realizing higher-quality sharing and opening of power data assets.

References

1. General Office of the Ministry of Industry and Information Technology: Notice of the General Office of the Ministry of Industry and Information Technology on Printing and Distributing the "Guidelines for Classification and Grading of Industrial Data (Trial)", 8 March 2020. http://www.cac.gov.cn/2020-03/08/c_1585210563153197.htm
2. China Mobile Communications Co., Ltd.: China Mobile Big Data Security Management and Control Classification and Classification Implementation Guide, vol. 12, pp. 7–14 (2016)
3. Market Supervision and Administration of Zhejiang Province: Digital reform—guidelines for public data classification and grading, vol. 7 pp. 8–10 (2021)
4. Institute of Policy and Economics, China Academy of Information and Communications Technology: Data value and data element market development report, vol. 5, pp. 29–38 (2021)
5. Xie, C.: Research on data rights confirmation in the context of big data, pp. 23–29. University of Posts and Telecommunications, Beijing, 1 May 2021
6. Wu, C.: Research on the legal issues of big data right confirmation, vol. 6, pp. 11–17. Jiangxi University of Finance and Economics (2018)
7. Xiong, Q., Tang, K.: Research progress on the rights, trading and pricing of data elements. Econ. Perspect. **2**, 144–153 (2021)
8. Li, Q.,Guo, C.: Rights confirmation of data resources: basic theory and application framework. China Popul. Resour. Environ. **30**(11), 212–214 (2020)

Big Data Technology in Real Estate Industry: Scenarios and Benefits

Xiangbo Zhu[1](✉) and Yi Li[2]

[1] Shenzhen Polytechnic, Shenzhen 518055, Guangdong, People's Republic of China
zxb@szpt.edu.cn
[2] Shenzhen Institute of Information Technology, Shenzhen 518172, Guangdong, People's Republic of China

Abstract. The real estate industry is the leading and pillar industry in the national economy, and plays an important role in promoting urban construction, boosting economic development. At same time, the real estate and its related industries have absorbed a large number of employed people. The advent of the era of big data has brought opportunities and challenges to the real estate industry, which require it to make the appropriate changes. By sorting out the application and analyzing some successful cases of big data technology in the real estate industry, this paper clarifies the common application scenarios and evaluating the performance of big data technology. In addition, the authors providing some suggestion for the future application of big data technology in the real estate industry.

Keywords: Real estate industry · Big data · Application scenario · Future path

1 Introduction

With the widely and continuous application of big data and related technologies, the real estate industry is inevitably involved in, Big data has gradually become the core resource and the basis for continuous innovation in the real estate industry. Some real estate companies have utilized big data technology, established the large information database, and successfully achieved the expansion of the business chain and area. While, due to resource deficiencies, some enterprises fail to effectively utilize big data, resulting in operational difficulties, and even forced to change careers and go bankrupt.

Real estate big data mainly includes real estate spatial location data, real estate market transaction data, and real estate-related macroeconomic, land, transportation, etc. Through the application of big data, real estate enterprises have created various scenarios such as business process optimization, business decision improvement, business boundary broadening, and integrated management. How to effectively collect and apply real estate big data and how can real estate companies effectively apply big data technology? How to price real estate big data and how to trade and share real estate big data? These questions are worthy of in-depth research.

B. Hu et al. (Eds.): BigData 2022, LNCS 13730, pp. 70–79, 2022.
https://doi.org/10.1007/978-3-031-23501-6_8

2 Literature Review

Big data technology is applied in all aspects of the real estate industry, including real estate development, real estate marketing, real estate evaluation, property management, real estate design, real estate centralized procurement and other fields. From theory and empirical aspect, scholars conducted in-depth research and achieved rich research results.

2.1 Real Estate Big Data Types

Spatial location data is an important type of real estate big data, such as latitude & longitude indicators, heat maps, and social network platform data. Juan. Y.K (2018) applied point of interest (POI) data tools to collect real estate information and analyze urban spatial layout [1]. Gautier. P.A (2017) collected the location information carried by users when visited websites, and applied spatial processing methods to explored the population distribution and spatial differences in different regions, the results could use to guide the job-housing balance in cities. In addition, some social network platforms also apply "user check-in functions" to obtain users' geographic location information, which can provide guidance for real estate development and consumption [2].

Behavioral trajectory data is another typical type of real estate big data, which can reflect the continuous spatial behavior of individuals. It mainly including mobile phone signaling data, traffic sensing data. Zhang Shun (2019) believed that mobile phone signaling data has strong temporal and spatial continuity, is relatively objective and comprehensive, and can be used to analyze urban spatial layout, population distribution, and it is helpful for real estate companies to select locations [3].

Huge consumption data is also an important source of real estate big data. Mo Tianquan (2013) proposed that based on massive information and big data technology, real estate companies can deeply mine customer needs. In addition, the real estate big data can more efficiently to develop new markets and determine the positioning standards of different regions [4]. Zhang Tao (2015) proposed to achieve precise positioning of target consumers based on consumers' consumption records, website browsing records, geographic location, and other information [5].

2.2 Real Estate Big Data Used in Development and Management Decisions

Big data technology has a wide range of applications in real estate development. Li Mengyang (2018) pointed out that big data can provide guidance for enterprise development decisions. For example, dig deep into government statistics to make macro market forecasts, and use consumption data to judge market prospects [6]. Chen Peng (2016) studied cases of real estate marketing and sort out the customer screening model, with which to standardize the classification of customers, and then achieve precision marketing. The way to achieve precision marketing is sufficiency using analysis technology on customer behavior data [7].

In addition, real estate big data can significantly improve the accuracy and efficiency of real estate valuation. Dong Qian (2014) believes that automatic real estate appraisal is an important direction of change in real estate appraisal. He discusses the technology of real estate automatic valuation [8]. Liu Changxian (2019) focused on land value assessment, emphasizing that big data technology can effectively assess land value, control investment risks, and improve assessment efficiency [9].

2.3 Real Estate Big Data Used in Real Estate Policy Formulation

Under the strategy of implementing big data and "Internet+", big data has become an important basis for the government to formulate housing policies. Accordingly, big data technology has become the important tools in exploring and formulate real estate prices [10–12]. Yu Weijun (2017) applied Baidu search data to establish a stepwise regression model for housing price prediction. Based on the data of Nanjing city, the housing price prediction model was optimized and the real-time prediction of the housing price trend was successfully realized [13]. Xue Bing (2018) took the relocation and reconstruction of Tiexi District in Shenyang city as the object, based on POI big data and GWR model, he explored the spatial differences of the impact of different factors on housing prices, providing a theoretical basis for the government to formulate reconstruction policies [14]. Cong Binghao (2019) studied the fluctuation of real estate prices and their influencing factors from the perspective of big data, and took Shiyan city as the object, to explore the possibility and method of city-specific policies [15].

Big data method has greatly changed the working ideas and decision-making methods of housing security. Wang Wenhua (2017) took Shenzhen city as an example, and based on the mobile phone signaling big data and activity trajectory big data, he finished the analysis of the population distribution characteristics and residential adaptability in various areas of the city. Through combined with the city's housing provident fund deposit big data, he provides guidelines for the layout and planning of affordable housing in Shenzhen [16]. In other research directions, Li Wei (2019) pointed out the application status of big data technology in the field of real estate archives management, and proposed that big data can effectively excavate the value of real estate archives, thereby improving the standardization and systematization of real estate archives management [17, 18].

2.4 Comments

In general, the existing research had covers almost all aspects of real estate big data [19–21]. However, the in-depth application still locked, and the comprehensive case study also need strengthen. The in-depth application of big data technology and methods will help to grasp the real estate market situation more accurately, formulate policies more scientifically, and better realize housing security.

3 Opportunities and Dilemmas of Real Estate Big Data Application

3.1 Opportunities

Big Data Improves the Decision-Making Efficiency. Big data is wildly applied in market trend judgment, internal management system reconstruction, overall structure optimization and many other aspects in real estate companies. First, in the context of big data, real estate companies can comprehensively analyze the construction and development process. From the process from project establishment, production, and construction to marketing and promotion, predict the future development status, big data can help the company in preventing potential risks, digging out possible industry business opportunities, and achieving benefits. Secondly, big data technology can be used to adjust complex management. Through rebuild the management system, big data help real estate enterprises maximize the efficiency of management. Third, big data improving the collecting and analyzing of market information. Real estate companies flexibly use big data platforms to collect data, optimize databases and expand data chains, and successfully realize data-based decision-making.

Big Data Improves the Efficiency of Marketing. In the era of big data, the information stock and information increment are huge. Search engines and social networks have become important channels for information acquisition and information sharing. First, in the real estate marketing, all work including products, channels, prices and customers are linked to big data. All kinds of data like business information, industry information, product experience, commodity transaction records are important to corporate decision making. Secondly, through collecting them from the search engines and social networks, we can make more accurate decision on real estate user positioning and product marketing. Similarly, with big data obtaining from survey and statistics, real estate companies can fully understand market information, master the business conditions and dynamics of competitors. Thirdly, by accumulating and mining consuming big data, real estate enterprises can further analyze the consumption behavior and value orientation of customers and achieve better service for consumers.

Big Data Supports the Benefit Management. Sell the right products to the right customers at the right time, with the right price and through the right channels, are the ultimate aim of enterprise. To achieve the goal of revenue management, demand forecasting, market segmentation and sensitivity analysis are three key links. First, in the demand forecasting link, enterprises collect product sales data and price data of market segments, establish mathematical models, and adapt scientific forecasting methods to forecast the potential demand, and then judge the product sales volume and sales price for a period in the future. Secondly, in the segment of the market, companies implement dynamic pricing and differential pricing for different market segments, and then maximize the revenue of each segment. Thirdly, in the sensitivity link, companies used big data on constructing the demand price elasticity analysis technology, and then achieve the control of factors that affecting price and demand in a forward-looking manner, which will better realize benefits.

3.2 Dilemmas

The establishment of a large database, the collection and analysis of data requires an abundant capital investment. At the same time, the processing of big data requires strong professional ability, and the operating cost is relatively high. In general, the in-depth application of big data in real estate enterprises faces the following challenges.

Big Data Analysis Technology is Relatively Lacking. In the era of big data, real estate companies can obtain massive data based on search engines, social networks and e-commerce platforms. These unstructured data have larger capacity, faster update frequency, and more complicated. The collection and analysis of real estate big data puts forward higher requirements to real estate companies. At present, big data technology specially developed for real estate enterprises is not perfect, and data management and deep mining technology also need to be overcome.

The Security of Big Data is Difficult to Guarantee. An important issue facing the era of big data is data privacy. The security of real estate big data involves not only technical aspects, but also social and ethical issues. In the process of collecting big data by real estate companies, reports that violate privacy regulations or data ethics appear from time to time. During vigorously promoting the application of real estate big data, attention should be paid to ensuring the data security. Maybe we can achieve the goal through standardized operation and supervision mechanisms.

The Real Estate Big Data Sharing and Transaction Mechanism Is Not Perfect. Due to the particularity of the real estate industry, real estate data resources is in the hands of government departments, financial institutions and internet companies, while real estate companies themselves have relatively little market data. How real estate companies can effectively obtain the right to use big data, and how to achieve reasonable sharing and trading of real estate big data are all important challenges in the development of real estate big data.

4 Innovative Scenarios for the Application of Big Data

4.1 "Internet+" Real Estate Marketing

Some companies use social platforms to ultimately facilitate real estate transactions through a combination of traffic, data, and multimedia platforms. Social tools such as cloud communities, WeChat marketing, Weibo marketing, APP message push, and Tiktok live broadcast are all used to attract buyers. For example, in 2020, China Evergrande Group launched "Hengfangtong", an online marketing tool that has won the trust of customers and achieved good sales performance. It has successfully realized the advantages of "real estate information disclosure (real estate information disclosure, clear price)" and "enhancing the convenience of house purchase (house purchase cost and time cost reducing)".

In 2021, China Country Garden Group build "Phoenix Cloud" as a digital marketing platform operating in line. The platform includes WeChat mini-programs and PC terminals, and meet five service functions. It builds a one-stop full-cycle home purchase service system for customers, digitize all business processes, and online modules of house search, house inspection, house selection, house purchase.

4.2 Personalization of Real Estate

With the help of the Internet and big data technology, real estate companies can directly connect to consumer demand and can achieve the customized of real estate design and decoration. In the process of real estate purchase, decoration and property management, consumers can fully express their wishes and highly participate in. In 2021, China Gemdale Group build a customized model of "pre-design, variable apartment type, customized hardcover, and personalized softcover pilot" to meet the individual needs of residents.

First, provide a product named "Golden Land Flexible Smart Home" to meet the flexibility of the apartment type. Customers can buy a suite to realize "change the type of apartment without changing the house". Based on the concept of "variable apartment type throughout the life cycle", the housing meet the demand of customers at different stages of life. Secondly, offered a product called "custom hardcover". By providing optional customized packages of "three styles and six modules", it can meet the customization requirements. Through offering "house type, style, grade, color, storage, kitchen, bathroom, performance, and soft decoration" Decoration service, successfully meet the customization.

4.3 Digital Property Services

Digitalization has become an important reform direction for property services. In 2021, ten departments including the Ministry of Housing and Urban-Rural Development of the People's Republic of China jointly issued the "Notice on Strengthening and Improving Residential Property Management" to encourage property service-related companies to build smart property service platforms.

In the era of big data, various high-tech smart devices have flood the property community, including multi-functional humanoid robots such as cleaning robots, security patrol aircraft quality inspection robots, distribution robots, computer room duty robots, parking robots, environmental monitoring robots and intelligent visitor robots. At the same time, the application of new electronic contract technology has become another breakthrough in the digitization of property service companies. Through the application of blockchain, cloud computing and other technologies, online services between owners and property companies are realized. For example, OCT Property, relying on digital technology, has built three major platforms including "OCT Smart Management Platform, 3D IoT Management Platform, and Super Housekeeper" to realize intelligent space service scenarios.

4.4 Upstream and Downstream Integration

Based on big data technology, real estate companies can open upstream and downstream businesses and achieve new business growth. Some leading real estate companies have achieved full coverage of industrial chains such as construction material procurement, financial insurance, construction industrial parks, real estate advertising, and real estate digital technology. In 2017, China Vanke Group and related companies jointly established the industry-wide B2B building materials procurement and trading platform

"Caizhu", which adopts the business logic of Tmall and Taobao and is positioned as "providing matching services". As of 2020, the platform has accumulated 1,085 registered customers, 46 real estate developers with sales of over 10 billion yuan, and a total turnover of more than 500 billion yuan.

5 Application Path of Big Data in Real Estate Industry

5.1 The Main Characteristics of Big Data Empowerment

Transparency. Transparency refers to the integration of various data, such as business data, IoT data, etc., into a unified platform under the means of big data technology. By successfully apply big data, enterprises can gain insight into all business conditions, and then form a smart enterprise under comprehensive perception. Digital business system. Through the overall business linkage and data integration at the enterprise level, enterprise achieved the overall strategic planning management at the operation level, and then the core business lines (including finance, cost, recruitment, planning, marketing, etc.) are integrally designed. Through the combination of data chain and the business chain, the real estate can achieve transparency of enterprise operation and market.

Predictable. Based on comprehensive perception of big data, artificial intelligence technology and big data algorithm are used to analyze the whole business process and value extension chain of real estate enterprises, and then achieve the real-time business simulation and prediction. From one aspect, comprehensive digitization, artificial intelligence, machine learning and other technologies can fit the key points of the business and provide information support. For example, applying a large amount of customer information to determine the preferences and needs of a particular customer. From another aspect, the application of big data technology to analyze business data and financial data, predict the possible cash flow of enterprises, and provide support for enterprise investment decisions.

Agility. With the support of high-performance IT equipment, artificial intelligence, cloud computing and the Internet of Things, real estate companies can quickly respond to changes. In the era of big data, real estate companies, as complex organizations, can self-adjust according to changes in internal and external environments, self-optimization, self-adaptation, and self-growth capabilities.

5.2 Feasible Paths for Big Data to Empower the Real Estate Industry

Build a Unified Operation and Decision-Making System. In order to make full and effective use of big data, real estate companies should establish standardized business standards. In the aspect of horizontally, real estate companies should open the data and processes of various professional departments. In the aspect of vertically, real estate companies should form closed loops such as assessment, management and control, and budget. Through build a unified a decision-making analysis platform, real estate companies can ensure that each business Changes in the links will fed back in real time.

At the same time, the application of big data to overall control the capital resources of the enterprise, integrate capital, cost, sales, progress, and contracts, real estate companies can realize the dynamic management of capital throughout the project cycle. Again, big data application can improve diversify business management. Through big data tools, centralized management and control methods are scientifically applied to achieve budget management, assessment management, human management, procurement management, investment management and other needs.

Diversified Construction of Marketing Channels. Real estate enterprises should make full use of search engines, social networks, live broadcast platforms, mobile APPs, and other channels to achieve service and marketing drainage, and better communicate with customers directly. At the same time, real estate enterprises make a diversification use of marketing techniques, such as virtual reality (VR) to provide viewing services. Real estate virtual reality allows customers to complete a home sale by being in it. Technology as artificial intelligence real estate consultants (Chatbot) and online intelligent question and answer robots are also highly recommending.

Applying Cloud Computing and Block Chain to Reconstruct Transactions. Real estate companies should choose different cloud computing solutions according to the needs of different departments and business lines. For example, the sales function can be customized and developed by adopting the PaaS method. Correspondingly, the financial system and human resource system can be completed by the relatively mature SaaS method. On the blockchain, transaction transparency is increased and transaction time is reduced through blockchain technology.

6 Conclusion and Suggestion

Embracing the Internet and big data is an important direction for real estate companies. How to find a model that suits the company's own advantages, how to use the internet to connect with upstream and downstream industries, and how to use big data technology to improve operational capabilities are questions that real estate companies need to answer. Based on the research of domestic and foreign scholars on real estate big data, combined with the practice application in real estate, this paper extracts the application scenarios of big data in the real estate industry and proposes the application path of big data in the real estate industry.

Facing the era of big data, real estate industry should strive to embrace big data and achieve better development. First, real estate enterprises should improve big data collection and analysis technology, and better build real estate big data. Secondly, relevant government departments should pay attention to the governance and ensure the safe operation of real estate big data, through standardized operation and supervision mechanisms. Third, all participants should actively build a real estate big data trading market. Build a data transaction mechanism and pricing model between big data owners, realize the value of data assets, and promote the innovative application.

Acknowledgements. This research was supported by Ministry of Education Humanities and Social Science Research Youth Foundation (No. 19YJC630239), Humanities and Social Sciences Annual Project of Shenzhen (No. SZ2022C020), Shenzhen Educational Science Planning Annual Project (No. Ybzz20008), Humanities and Social Sciences Youth Project of Hubei Provincial Department of Education (No. 17Q056).

References

1. Juan. Y.K., Chen, H.H., Chi. H.Y.: Developing and evaluation a virtual reality-based navigation system for pre-sale housing. Appl. Sci. **8**(6), 952–957 (2018)
2. Gautier. P.A., Siegman. A., Van Vuuren. A.: Real estate agent commission structure and sales performance, working paper in economics (2017)
3. Shun, Z., Nana, Y.: Research on the spatial distribution characteristics of Nanjing employment and housing under multi-source big data- based on the coupling analysis of mobile phone signaling data and Urban Points of Interest. In: Vibrant Urban and Rural Beautiful Habitat - 2019 China Urban Planning Annual Conference Proceedings, vol. 5, pp. 1098–1108 (2019). (in Chinese)
4. Tianquan, M.: China real estate market analysis and big data application. China Real Est. Valuation Broker. **6**, 18–22 (2013)
5. Tao, Z.: Research on the precise placement strategy of real estate advertisements based on the background of big Data. Chongqing, Southwest University, pp. 30–37 (2015) (in Chinese)
6. Mengyang, L.: Research on the application of big data in investment decision of real estate project development. Chongqing, Chongqing University, pp. 20–26 (2018). (in Chinese)
7. Peng. C.: The Research on the precision marketing strategy of Dalian a real estate company base on big data Dalian, Dalian Maritime University, pp. 40–47 (2016). (in Chinese)
8. Qian, D., Nana, S., Wei, L.: Real estate price prediction based on web search data. Stat. Res. **31**(10), 81–88 (2014). (in Chinese)
9. Changxian. L.: Study on the batch evaluation of commercial real estate price based on internet data. Chongqing, Chongqing University, pp. 34–39 (2019). (in Chinese)
10. Jianjian. Y.: The change and influence of "Internet + Virtual Reality" on real estate sales model Beijing, Beijing University of Posts and Telecommunications, pp. 60–66 (2019). (in Chinese)
11. Shengsheng, T.: Research on "Internet+" real estate marketing model and strategy. Hangzhou, Zhejiang University of Technology, pp. 57–66 (2018). (in Chinese)
12. Jing, L.: A summary of the development of big data in the real estate field. J. Henan College Fin. Tax. **34**(3) 60–64 (2020). (in Chinese)
13. Weijun. Y.: Research on Internet Marketing model of real estate agency industry under the background of big data. Tianjin, Hebei University of Technology, pp. 24–30 (2018). (in Chinese)
14. Bing, X., Xiao, X., Jingzhong, L., Jiang, L., Xiao, X.: POI-based analysis on retail's spatial hot blocks at a city level: a case study of Shenyang, China (in Chinese). Econ. Geogr. **38**(05), 36–43 (2018)
15. Binghao, C.: The influence of Internet finance development on China's real estate price fluctuation Jinan, Shandong University, pp. 33–38 (2019). (in Chinese)
16. Wenhua, W., Dongli, D.: Exploration and thinking on the use of big data to strengthen the supervision of real estate appraisal industry. China Real Estate Appraisal Agency **5**, 55–59 (2017). (in Chinese)
17. Wei, L.: Opportunities and challenges of the real estate industry in the era of big data. Chin. Foreign Entrepreneurs **19**, 80–84 (2019). (in Chinese)

18. Nan, L., Duo, X., Yun, C.: Research on the challenges and countermeasures of real estate information service under big data. Constr. Econ. **38**(2) 77–81 (2017). (in Chinese)
19. Peng, L., Xin, L.: Assisting real estate macro decision-making with big data. Macroecon. Manag. **4**, 34–36(2017). (in Chinese)
20. Sarathy, P.S.: TQM practice in real estate industry using AHP. Qual. Quant. **47**(4), 2049–2063 (2013)
21. Anselin, L.: Under the hood issues in the specification and interpretation of spatial regression models. Agric. Econ. **27**(3), 247–267 (2015)

Copyright and AI: Are Extant Laws Adequate?

Jingyi Cui(✉)

Jiangxi University of Finance and Economics, Nanchang, China
2202020866@stu.jxufe.edu.cn

Abstract. On September 22, 2022, the UK government released "Examining patent applications relating to artificial intelligence (AI) inventions: The Guidance", which clarified nine implementation guidelines. It specifically states that in the UK, AI inventions in all technical fields are patentable, and AI inventions are not excluded when the tasks or processes performed by AI inventions reveal a technical contribution to known technology and are patentable qualifications. However, does the existing copyright system provide sufficient protection for AI's creative innovations? Starting from the existing legal provisions, case law and doctrines, this paper analyzes the different attitudes of the EU, the US, Australia, the UK and China towards AI output works, and examines the inadequacy of the current copyright system in protecting AI's creative innovations by combining cases and survey reports. Based on the above analysis, at the end of the article, suggestions are made for the improvement of copyright law for AI protection, and conclusions are finally drawn.

Keywords: AI · Copyright · Patent

1 Introduction

In recent years, with the development of AI technology, AI's artistic creations have become more frequent, and today AI creations cover almost the entire range of subjects listed in Article 2(1) of the Berne Convention. There is no doubt that technological change promotes institutional change. Scholars began intentionally studying the legal issues related to computer-generated works as early as the 1960s, and with the development of AI technology cited in recent years, it is easy to see that the copyright law system, the patent law system, and the entire intellectual property legal system are responding accordingly. On September 22, 2022, the UK government released "Examining patent applications relating to artificial intelligence (AI) inventions: The Guidance", which clarified nine implementation guidelines. It specifically states that in the UK, AI inventions in all technical fields are patentable, and AI inventions are not excluded when the tasks or processes performed by AI inventions reveal a technical contribution to known technology and are patentable qualifications [1]. Encouraging AI technological innovation, promoting its use for the public good and maintaining the central role of intellectual property rights in promoting human creativity and innovation are consensus issues, and in the process of technological development, countries are seeking better ways to protect and promote the use of AI in innovation and creativity. However, does

B. Hu et al. (Eds.): BigData 2022, LNCS 13730, pp. 80–87, 2022.
https://doi.org/10.1007/978-3-031-23501-6_9

the existing copyright system provide sufficient protection for AI's creative innovations? Starting from the existing legal provisions, case law and doctrines, this paper analyzes the different attitudes of the EU, the US, Australia, the UK and China towards AI output works, and examines the inadequacy of the current copyright system in protecting AI's creative innovations by combining cases and survey reports. Based on the above analysis, at the end of the article, suggestions are made for the improvement of copyright law for AI protection, and conclusions are finally drawn.

First, the subject qualification of artificial intelligence. According to China's copyright law, authors include natural person authors and legal person authors; The former refers to the citizen who created the work, while the latter refers to that when the work is hosted by a legal person, created on behalf of the will of the legal person, and the legal person assumes responsibility, the legal person is regarded as the author. Recognizing that AI is the author, in fact, means creating a new independent legal subject in the copyright law, which will encounter great legal and ethical obstacles, and may be difficult to achieve for a long time.

The second is the work qualification of artificial intelligence products. The basic theory of copyright law holds that works should be human intellectual achievements, and only human intellectual activities can be called creation. Before the issue of the copyright of artificial intelligence products attracted wide attention, the legal community had discussed whether the content produced by animals could constitute a work. For example, in the United States, a black macaque used a photographer's camera to take several self portraits, and its copyright issue even led to two lawsuits. To this end, the United States Copyright Office has also issued relevant documents, emphasizing that only works created by human beings can be protected. Some scholars believe that artificial intelligence products are not the intellectual achievements of human authors, so they do not constitute works. Some scholars also believe that artificial intelligence products are the results of software generated by works designed by human authors. In fact, they are the intellectual results of human-computer cooperation and do not violate the personality basis of copyright law.

Third, the ownership of artificial intelligence products. At present, there are mainly 3 schemes proposed. The first solution is to recognize that AI products are works, but do not protect them and put them into the public domain. The main reason is that the legislative purpose of the copyright law is to encourage the creation and dissemination of works, while machines need not be encouraged. The second solution is to create a new neighboring right system to distinguish the works generated by artificial intelligence from those created by human beings. The third scheme is to make appropriate legal arrangements through legal interpretation under the framework of the current copyright law. As for whether the copyright belongs to the owner, developer or user of artificial intelligence, there is no consensus.

2 Protection of AI Creation by Existing Copyright System

Not all works of art are entitled to copyright protection. Most jurisdictions require three elements of copyright: fixation, originality, and artificial creativity. Original literary, dramatic, artistic, and musical works, as well as film, audio, video, broadcast, and published copies, are all protected by traditional copyright. Artificial creativity, regardless of how it manifests, is the most relevant study issue among them. In the invention and innovation of AI, these three basic prerequisites have been challenged, resulting in dramatically diverse ideas and precedents in different jurisdictions.

The US Patent Office has clearly stated that "copyrighted works must be created by human beings". Meanwhile, in case law, courts in the United States and Australia have emphasized the importance of human writers, stating that computer-generated information "have no authors and are not protected by copyright [2]". In the context of EU copyright law, there are no positive restrictions on this subject, but Article 1 of the Term Directive [3] can be considered a broad definition. It has become a consensus to comprehend works in copyright law as "a literary or artistic work within the meaning of Art. 2 of the Berne Convention, [4]". Similarly, the European Court of Justice has used this provision to set precedents and has integrated it into the TRIPS and WCT agreements. It is not difficult to recognize a common idea under EU law, that is, "the author's intellectual production," from the cases of Infopaq v. Danske, Levola Hengelo [5], Funke Medien, and D Brompton Bicycle Ltd v. Chedech [7].

Unlike the European Union, the United States and Australia, Britain is one of the few countries that protects computer-generated works. Rather than evolving through case law, the UK's Copyright, Designs and Patents Act 1988 provides some legal authority for authorship and computer-generated works. Section 1(1) states that "copyright exists in… Original… Property rights in works of art." This "originality" is normally tested in two ways in case law practice: the work must not be a copy, and it must be the result of the author's "talent, labor, and judgement". In light of the UK's former status as a Member of Europe, "talent, Labor, and judgement" has recently been construed to include the EU's creative criterion for "intellectual production" that "reflects the personality of the author [7]". he terms "computer-generated" is defined under Article 178 of the Act as a work "produced by a computer in the absence of a human author." Section 9(3) then states that a computer-generated work can be copyrighted and that the author "must be presumed to have made the essential arrangements for its development in the instance of a computer-generated literary, dramatic, musical, or aesthetic work." "The person who creates the required arrangements for the creation of the work" is defined as an author of a computer-generated work (CGW). The term of protection begins on the day of completion and ends 50 years afterwards. Neighboring rights, in addition to original literary, dramatic, musical, and aesthetic works, are also protected, and these rights objects do not have to be original. However, this lowered level of protection comes with fewer and shorter rights. For example, recording protection applies limited to a single recording of a song and lasts for 70 years from the date of its composition. Related rights apply if the AI-generated work fits into one of these categories.

Similarly, Chinese court, seeking a positive breakthrough in the clash between AI and copyright protection, issued the first legal confirmation in China that works generated by AI can be protected by copyright [8]. Shenzhen's Nanshan District Court acknowledged the selection and arrangement of human producers to participate in the creation of relevant output. The court decided that Dreamwriter, an intelligent writing computer software, created output that met the standards of a written work, and hence was protected by China's copyright law [5].

The plaintiff, Shenzhen Tencent Computer System Co., LTD. (Tencent), has used its affiliated company to manage its creative personnel to build an intelligent writing aid named Dreamwriter since 2015. Tencent produced a financial commentary automatically written by Dreamwriter on August 20, 2018, headlined "Noon Review: Shanghai Composite Index increased 0.11 percent to 2671.93 points, led by telecom operations, oil mining, and other sectors," for the first time. At the end of its lunchtime stock comments, Tencent added "[T]his piece was automatically created by Tencent robot Dreamwriter," indicating that the article was the plaintiff's legal labour. On the same day that the stock afternoon comment was published, the defendant, Shanghai Invensys Technology Co., LTD. (Invensys), published a review on the defendant's "Wangdaizhijia" website without the plaintiff's authorization. Tencent filed a lawsuit against Winxun in Shenzhen's Nanshan District Court, alleging that Winxun's actions infringed on its right of information network communication and constituted unfair competition. The Nanshan District Court of Shenzhen city issued a partial decision in favour of the plaintiff on December 24, 2019, classifying the article as a legal work owned by the plaintiff and the plaintiff's information network dissemination as an infringement. The Nanshan District court in Shenzhen approved copyright protection for Dreamwirter's automatically created articles, marking the first attempt in Chinese legal practise to provide copyright protection for AI invention and innovation. In the current human-centric world of modern copyright law, the most significant significance of the movie and Tencent instances is that they provide a copyright protection framework for ai-generated material. Based on China's court experience, artificial intelligence-generated content, at least some of it, can get copyright protection.

3 Inadequate Protection of AI Under Copyright Law

Over decades of evolution, copyright law has efficiently reacted to a variety of difficulties posed by technological advancement [9]. Clearly, the methods of the United States, Australia, and the European Union are too direct to disregard artificial intelligence's rights and interests in creating works. The fact that AI tools and interfaces are currently available to everyone under an open licence does not fully meet the requirements for encouraging AI technology innovation, promoting its use for the public good, and maintaining intellectual property's central role in promoting human creativity and innovation. Over decades of evolution, copyright law has efficiently reacted to a variety of difficulties posed by technological advancement. Clearly, the methods of the United States, Australia, and the European Union are too direct to disregard artificial intelligence's rights and interests in creating works. The fact that AI tools and interfaces are currently available to everyone under an open licence does not fully meet the requirements for protecting the implementation of AI in creativity and innovation. However, the attempts represented by Britain's Theory and China's Practice have shown numerous flaws, most notably in the form of ambiguous legal concepts and inadequate protection.

The UK IPO issued a request for comments on AI and intellectual property from September 7 to November 30, 2020, and the results demonstrate that the current framework can meet future challenges in many areas. Ordinary copyright law protects works generated by humans employing AI as a tool [10]. However, there is no doubt that the current copyright structure is not well suited to the use of AI in creative creation. Existing methods for safeguarding AI works are insufficient. The law on computer-generated works is now uncertain due to a lack of understanding of what originality means in respect to such works and case law tying originality to human creativity, and the UK Intellectual Property Office has re-issued a request for comments on October 29, 2021.

In the last request for comments, it's seems commonly acknowledged that "general artificial intelligence" has not yet been achieved. And, under Section 9(3) of the Copyright, Designs and Patents Act (CDPA) 1988, ai-generated works already fall under the definition of computer-generated works, and this protection should be maintained. However, the United Kingdom's Copyright, Designs and Patents Act 1988 has a contradictory stance on human authorship (and, by extension, originality), stating that computer-generated work has "no author" (Section178), while simultaneously defining "who the author should be" (Section 9)(3). This has, understandably, caused some consternation. If there is "no author" under article 178, according to Bonadio, this "obviously provides an exemption to the criterion of originality" [11]. Dickinson, on the other hand, claims that authorship nomination (3) in article 9 simply transfers the originality requirement to that individual. "Whether the accredited author (i.e., the person who makes the essential preparations for the creation of a work by computer) has used his/her skill, labor, and judgement in that creation" [12], she believes the statute requires. She also points out that this method is consistent with Nova Productions V Mazooma Games' "extremely restricted rationale. "The majority of articles on this topic have referenced to the case of Express Newspapers v. Liverpool Echo [14], which established a distinction between "computer created" and "computer assisted" work [15]. The court determined that "in computer-aided labor, software is merely a tool for making the final output, and hence the copyright belongs to the person who utilizes the software [16]" as Dorotheou phrased it. This distinction clarifies the law: if the work is computer-generated, it is exempt from authorship and originality standards; nevertheless, if it is just computer-aided, the human author must look original. Because copyright existing in both circumstances, Section 9(3) should not be seen as introducing "writers" for the purpose of originality assessment, but rather as determining who should profit from copyright, causing misunderstanding regarding the statute's meaning. In extreme circumstances, AI systems prevent users from making any meaningful decisions other than pressing a few buttons [17].

At the same time, the present "author's own intellectual production" originality criteria does not support copyright protection for AI works. It is preferable to develop new methods for determining if copyright protection exists based on the creative output or creative process. Human innovation should take precedence over machine creativity, according to copyright. The copyright framework in the United Kingdom is critical for keeping human incentives for production and distribution alive. Ordinary copyright law is usually assumed to protect works generated by humans utilising AI as tools. However, most individuals believe that content created solely by AI (without human creative expression) either does not qualify for copyright protection at all or should be protected differently. Although certain jurisdictions currently protect certain types of computer-generated works, especially given the explicit and intentional extension of copyright regimes to protect such works, there can be no guarantee that the prior explicit protection will cover all types of computer-generated works. New forms of computer-generated work continue to emerge, such as robotics, machine learning, artificial intelligence, and the Internet of Things in general. As a result, any AI-assisted output generated by such systems may not qualify as a "work," exposing the flaws in the current copyright legal framework.

4 Suggestions on Perfecting the Copyright Law Protection of AI

For years, the topic of copyright protection for artificial intelligence-generated material has been high on political agendas around the world, igniting one of the most complex, acrimonious, and interesting debates in modern copyright law. The existing human-centered copyright law framework is the main impediment to copyright protection for ai-generated outputs in most jurisdictions. In the face of artificial intelligence's problems, and as we discuss the emergence of new modes of communication, we must carefully consider "whether a new expression of rights is required, or whether the old expression is still applicable and only flexible interpretation is required to adapt to these changed conditions?".

First, the definition of artificial intelligence creation and innovation must be clarified. When it comes to copyright protection for AI, the first question to ask is whether the product generated by AI is original. When coming up with an innovative question, it's best to be objective. In the creative process, the automatic operation of ARTIFICIAL intelligence should not be an impediment to identifying uniqueness. This objective method also corresponds to current development trends. Copyright protection of AI output solves the difficulty faced by the original machine author by employing an objective approach.

Second, for AI-assisted output, the key is to determine if there are sufficient human aspects to construct AI software without input keywords, or whether output provided by input keywords alone does not constitute sufficient human intervention, while advanced selection and arrangement, there can be enough human intervention to make the AI-generated output copyrighted, such as the entry of data kinds, the setting of trigger conditions, and the selection of templates and corporates. The evaluation of human intervention should not be restricted to the precise short amount of time in which outputs are created; rather, it should include the choices and arrangements made ahead of time.

Finally, as artists make more extensive use of AI and machines grow more capable of making creative works, the line between human-created art and computer-created art will become increasingly blurred. The vast advancements in computing, as well as the sheer amount of processing power accessible, are likely to render the distinction meaningless. Machines get better at impersonating humans when you give them the capacity to learn styles from vast content collections. We may soon be unable to distinguish between human-generated and machine-generated content if computational power is available. We're not quite there yet, but once we are, we'll need to think about new safeguards for upcoming works made solely by intelligent algorithms. The current state of copyright has made numerous concessions to the growth of ARTIFICIAL intelligence, gradually moving away from the original norms that rewarded skill, work, and effort. However, perhaps we should make an exception to this trend when it comes to complicated AI breakthroughs [18].

5 Conclusion

As what have shown above, it's appropriate for AI to be protected by copyright, but it needs to be considered in different scenarios. We need to provide more extensive and flawless AI protection in order to foster innovation and drive economic growth. Copyright protection for AI-assisted production is currently in place in the UK and China, and countries are deliberately broadening the definitions of "original" and "author." The conclusion, however, is that AI protection is insufficient, which is clear. In fact, in the United Kingdom and China, there has yet to be a conclusion on whether artificial intelligence can be considered a qualified copyright owner, leading to misunderstanding and incorrect conduct. This could give humans the ability to infringe on the rights of AI's creations, tools, and interfaces. Our system of copyright law will not be able to manage and safeguard AI when it reaches the stage where it can export creations on its own. As a result, we should clarify the concepts of "originality" and "author" within the context of existing copyright, as well as the appropriate relaxation limit, the output of the auxiliary for artificial intelligence while also strictly grasping the limits of human intervention, and in the face of advanced artificial intelligence work in the future, looking for a new system of protection, may be the best way.

At the same time, there will be many legal problems in the subsequent development of AI products. For example, the infringement of AI products. In the process of "machine learning", AI needs to use a large number of existing works. For example, the robot "Xiaobing" created a collection of poems after learning many modern poems, of which some works are still under copyright protection. Then, does the commercial use of his work without the authorization of the author constitute infringement? It is generally believed that in order to promote the development of artificial intelligence, the use of other people's works in the process of "machine learning" should be treated as an exception. This will also be a major issue that we need to discuss and study.

References

1. GOV.UK.: Artificial Intelligence and Intellectual Property: copyright and patents (2021a). [online] https://www.gov.uk/government/consultations/artificial-intelligence-and-ip-copyright-and-patents/artificial-intelligence-and-intellectual-property-copyright-and-patents. Accessed 6 Jan 2022
2. GOV.UK.: Government response to call for views on artificial intelligence and intellectual property (2021b). https://www.gov.uk/government/consultations/artificial-intelligence-and-intellectual-property-call-for-views/government-response-to-call-for-views-on-artificial-intelligence-and-intellectual-property. Accessed 6 Jan 2022
3. Guadamuz, A.: Artificial intelligence and copyright. [online] Wipo.int (2017). https://www.wipo.int/wipo_magazine/en/2017/05/article_0003.html. Accessed 6 Jan 2022
4. Hugenholtz, P.B., Quintais, J.P.: Copyright and artificial creation: does EU copyright law protect AI-assisted output? IIC – Int. Rev. Intellect. Prop. Compet. Law (2021)
5. Williams, B.: Painting by numbers: copyright protection and AI-generated art. Eur. Intellect. Prop. Rev. (2021)
6. Balganesh, S.: Causing Copyright. Columbia Law Rev. (2017)
7. Acohs Pty Ltd v. Ucorp Pty Ltd [2012] FCAFC 16 (Aus)

8. Dorotheou, E.: Reap the benefits and avoid the legal uncertainty: who owns the creations of artificial intelligence? Comput. Telecommun. Law Rev. (2015)
9. Bonadio, E., McDonagh, L.: Artificial intelligence as producer and consumer of copyright works: evaluating the consequences of algorithmic creativity. Intellect. Prop. Q. (2020)
10. Dickenson, J., Morgan, A., Clark, B.: Creative machines: ownership of copyright in content created by artificial intelligence applications. Eur. Intellect. Prop. Rev. (2017)
11. Shenzhen Tencent Computer System Co Ltd., v Shanghai Yingxun Technology Co Ltd. Yue 0305 Min Chu No. 14010 (Nanshan District Court of Shenzhen) (Tencent) (2019)
12. Zhang, W.: The first case in the field of AI writing was decided, the court confirmed for the first time that work generated by AI was original and protected by the copyright law. Legal Daily **8** (Beijing, 8 January 2020)

Author Index

Printed in the United States
by Baker & Taylor Publisher Services